HOW TO BUILD
GREENHOUSES
GARDEN SHELTERS & SHEDS

HOW TO BUILD

GREEN

GARDEN SHELTERS

A POPULAR SCIENCE BOOK

HOUSES
& SHEDS

by Thomas H. Jones

Drawings by Vantage Art, Inc.

POPULAR SCIENCE ● HARPER & ROW

New York, Evanston, San Francisco, London

Library of Congress Catalog Card Number: 77-26476
ISBN: 0-06-012218-8

Manufactured in the United States of America

To Carolyn

Contents

Preface

In this book, you'll find a complete guide for design and construction of home greenhouses. You'll also find a substantial section with design ideas and some construction details on a wide range of backyard shelters, including gazebos, garden work centers, storage sheds, and more.

Home greenhouses, the forerunners of modern solar houses, satisfy many needs. They provide a place to cultivate and show plants—with flowers blooming year-round. And they allow you to grow vegetables throughout the winter. Home greenhouses need not be confined to estates and large suburban yards. City dwellers can have greenhouses too—on rooftops or even in cellars.

Whether you are thinking about buying a kit or building your own greenhouse from scratch, there is a lot more to setting up a functioning greenhouse than just putting a layer of transparent material between your plants and the out-of-doors. This book covers the basics, including the different kinds and styles of greenhouses, practical considerations for selecting the site, tips on zoning and construction permits, and means of climate control inside the greenhouse.

The book then takes up the ways of getting the greenhouse you want. You'll find a survey of available kits, and details for do-it-yourself design and construction, including the basics for making foundations. Producing the climate you want in a greenhouse is mainly a matter of the right equipment, such as heaters and heating plants, ventilation systems, humidifiers, and coolers. Here, equipment and options are described in detail. Step-by-step greenhouse heating requirement calculations are given, followed by advice on how to conserve heat.

As to other backyard structures, there can be many similarities in materials and construction techniques, as well as an overlap in uses. And foundations for these other structures may be quite like those for some greenhouses. A gazebo is certainly a shelter, and you could easily store tools underneath one, or enclose part of the gazebo for a storage shed or poolside dressing room. Garden work centers and tool sheds are often combined. Small storage sheds store not only tools and items that would normally be in the attic, garage, or basement. Sheds are also used for work centers, workshops, home offices, and guest rooms. Some sheds have enclosed porches for outdoor entertaining, and some little barns are actually used as horse stables. In short, the difference between a gazebo and a shelter can be quite arbitrary: Your shelter may be my gazebo.

Writing a book like this is not a solo effort. I received a lot of help from many companies in the greenhouse and shelter business. They provided information, illustrations, and suggestions. Special thanks go to architect Adolf deRoy Mark for material on urban greenhouses.

Thomas H. Jones

PART *1*

Greenhouses

A free-standing gable greenhouse can be located anywhere on the property to get the best combination of sun, wind protection, summer shade, and convenience. It can also be attached at one end to a house. Shown is the new Janco glass-to-ground Camelia model. (*Courtesy J. A. Nearing Co., Inc.*)

1

Greenhouse Basics

Home greenhouses serve many purposes. Some greenhouse owners cultivate and show only rare and exotic plants. Other people are most concerned with having flowers in bloom throughout the winter. Still other gardeners may use a greenhouse primarily for starting plants early in the spring and, later, for extending the growing season far into winter. Vine-fresh tomatoes and peppers and other vegetables can be grown in a greenhouse all winter long. A greenhouse can serve just as a pleasant, quiet green place where you can relax after a hard day's work. And if it adjoins a dining room or living room, it can add an elegant touch to the daily routine as well as to parties.

There are many styles of greenhouses for the home, and greenhouses can be made of various combinations of materials. There are advantages and disadvantages to each style and material.

The most common greenhouse styles are the gable and the lean-to. The gable greenhouse, also called the even-span because it has two equal-sloping roofs, can be located anywhere on your property, or it can be attached to your house by one gable end, allowing you direct access to the greenhouse from indoors. The free-standing-gable greenhouse can be sited for best exposure to winter sunlight, drainage, and wind protection—definite advantages.

The lean-to greenhouse is the most popular style. It must be attached to your house or some other structure. But this severely limits the places you can put it if you want optimum sunlight exposure. Set

The lean-to greenhouse design can provide access from inside the house. When the greenhouse is attached to a house, heating costs are less than for a free-standing greenhouse of the same size. Lean-tos require foundations to prevent separation from the main building when the ground heaves. Shown is the Straight-Eave Orlyt 9 by Lord & Burnham.

City greenhouse designs require ingenuity. The cellar greenhouse, at right, has a wired-glass cover mounted in steel plate for safety and security. (*Adolf deRoy Mark, Architect*)

The greenhouse at the extreme right was built over a cellar-level courtyard. It is two stories high and is designed primarily for plants with trailing foliage. (*Adolf deRoy Mark, Architect*)

This rooftop lean-to greenhouse by Vegetable Factory has a two-layer flat fiberglass-reinforced plastic cover bonded to aluminum frames to minimize heat loss. (*Courtesy Vegetable Factory, Inc.*)

up against an opened wall of your house, it can provide an attractive and colorful extension to a room. A lean-to greenhouse costs less to heat than a free-standing greenhouse of the same size and construction because there is no heat loss through the greenhouse wall common to the house (assuming you are not running the greenhouse at a higher temperature than your home). It is also possible to heat a lean-to by extending your home heating system, or by just leaving the connecting door open, with the possible added assistance of a small fan to draw in the warm air. This can result in a significant reduction in annual operating expense, although it will add to your home heating cost.

A window greenhouse might be classed as a miniature lean-to. If you don't feel up to the responsibilities of a large greenhouse or if you just want to experiment before you drop a lot of money into a greenhouse, an inexpensive window greenhouse, either kit or home-built, is the way to start. Window greenhouses fit into or cover a window. Their size is limited by how far you can reach from inside the window, but there is still enough room for growing flowers and even tomatoes. Heat is supplied either by keeping the indoor window par-

BASIC GREENHOUSE STYLES

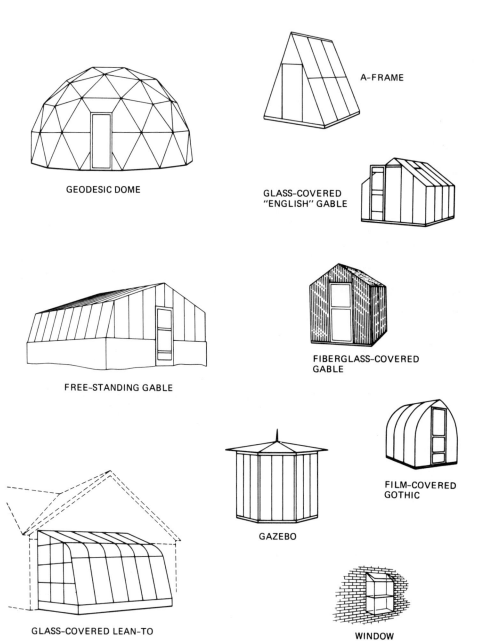

GEODESIC DOME

A-FRAME

GLASS-COVERED
"ENGLISH" GABLE

FREE-STANDING GABLE

FIBERGLASS-COVERED
GABLE

FILM-COVERED
GOTHIC

GLASS-COVERED LEAN-TO

GAZEBO

WINDOW

A window greenhouse is a good starter unit if you're unsure of greenhousing as a hobby. Cost is moderate. This Janco window greenhouse comes in several sizes, and two same-size units can be combined for an extra-wide window. (*Courtesy J. A. Nearing Co., Inc.*)

This tiny triangular window greenhouse has a tinted acrylic cover. (*Adolf deRoy Mark, Architect*)

tially open, or by placing a low-wattage resistance heater in the bottom of the greenhouse.

A greenhouse in the shape of a dome is inexpensive to build and excellent for small properties. Domes provide a lot of ceiling height, but they are prone to water leakage. A-frame greenhouses are the simplest to build, but except at ground level, growing space is limited in comparison to the area occupied by the greenhouse. The least expensive greenhouse is the pit greenhouse sunk into the ground with only the south-facing window in glass.

These two Greendome geodesic-dome greenhouses are connected by a sealed walkway. The frames are aluminum. The covers are film. (*Courtesy Dome East Corp.*)

Here is the interior of Vegetable Factory's free-standing greenhouse, in this case installed on a rooftop. Two-layer translucent acrylic and fiberglass reduces heating costs.

The materials used in the construction of a greenhouse have a greater influence on its overall appearance than does the style. If you want a greenhouse with crystal-clear sides and roof revealing a colorful profusion of flowers and vegetables and greenery on the inside, then your choices include wood or aluminum framing with a glass or *clear* plastic cover. These are the most expensive greenhouses to buy or build.

On the other hand, if you are more interested in a strictly functional greenhouse for raising vegetables or flowers, choose a rigid plastic or a flexible-film covered greenhouse. There is a lot of variety in plastic-covered greenhouses. Covers vary from almost clear down to something that appears to be opaque white from the outside. Rigid plastic-covered greenhouses cost less than glass-covered styles, and film-covered types have an even lower initial cost. Rigid plastic covers last for years, but not as long as glass. Most film covers do not last more than a year, but they are inexpensive to replace.

This does not mean that a strictly functional plastic-covered greenhouse cannot be beautiful inside. It will simply not look like a traditional greenhouse from the outside.

COVERS

The cover on a greenhouse functions 1) as a window to let sunlight into your plants, 2) as an enclosure to retain heat, and 3) as a barrier to keep snow, rain, hail, high winds, birds, small animals, and insects away from your plants. Each of these three types of covers—glass, rigid plastic, plastic film—has advantages.

Next in importance to siting your greenhouse is the selection of cover material. For the choice of cover material sets the requirements for the greenhouse frame and foundation. First, decide whether the cover will be permanently installed glass or rigid plastic, or a plastic film that will have to be replaced on a regular schedule.

Glass is the oldest and the traditional cover material. (In Britain, greenhouses are called glasshouses.) Glass is the heaviest cover material and requires a strong frame that will support the weight without bending. And glass requires more closely-spaced frame members because glass should be used in smaller pieces than is rigid plastic.

The Sun Glory A-frame greenhouse has adjustable side shades and glass-to-ground construction. It provides lots of growing space. (*Courtesy Sturdi-built Manufacturing Co.*)

The cover of a 6 × 8-foot, even-span greenhouse would weigh 370 pounds, if the glass were double-strength. A single-strength glass cover would weigh 275 pounds. The cover would weigh only 74 pounds if it were rigid plastic. And it would weigh only 5 pounds if it were 6-mil film.

Three kinds of rigid plastics are used for greenhouse covers—fiberglass, acrylic, and polycarbonate. Fiberglass is actually a panel of polyester resin reinforced with short pieces of glass fibers. It is made in flat and corrugated panels in many colors and grades. Not all varieties are suitable for greenhouse covers.

You cannot get clear transparent fiberglass; it is only translucent. Yet careful formulation of the polyester resin can produce a sheet that is almost transparent. Fiberglass weathers, becoming dirty-looking and less translucent over a period of 10 to 20 years, depending on weather conditions and exact resin formulation. The resulting appearance leaves something to be desired, and there's also a performance problem. The imbedded dirt reduces light transmission, and this affects plant growth. The weathering mechanism is erosion of the polyester surface which exposes the glass fibers. These fibers in turn collect wind-blown dust and dirt. You can extend the life of a fiberglass greenhouse cover by scrubbing it down. To improve resistance to weathering, some manufacturers coat the fiberglass panels with Tedlar, a DuPont polyvinyl fluoride.

Acrylic plastic, commonly known as Plexiglas, which is actually Rohm & Haas' trade name for the product, is a greenhouse cover material with many advantages. Compared to single-strength and double-strength glass, acrylic can be up to thirty times as impact-resistant in the same thicknesses. One variety is crystal clear, and it stays clear. Clear, tinted varieties are also available. Unlike glass, acrylic plastic is combustible, and its surface finish can be ruined by chemicals and abrasives in common glass cleaners.

Because of its strength, acrylic is generally used thinner than glass, and larger pieces can be used. But reinforcing bars must be added to prevent the larger flexible sheets from being blown in or out by high winds.

Polycarbonate—General Electric's Lexan—is for security glazing. Its brick-stopping properties are well known. However, it's not a good choice for greenhouse glazing because it weathers poorly, turning yellow after four years and milky white after seven. Both polycarbonate or acrylic windows (lites) can be used in place of tempered glass in doors.

Plastic film covers for greenhouses cost far less than glass or rigid plastic covers but they do not last very long. Three films are used—polyethylene, polyvinyl cloride (PVC or vinyl), and polyester (Mylar).

Polyethylene film is used in 4- and 6-mil thicknesses for greenhouse covers and is probably used to cover more commercial greenhouses than any other material. Ordinary vapor-barrier and dropcloth "polyfilm" sold at lumber yards and home centers should not be used because it seldom lasts three months in sunlight. Ultra-violet-inhibited polyethylene films specially formulated for greenhouse covers will last one to two years. The film, such as Monsanto's "602," comes in 100-foot rolls up to 40 feet wide. With it, you can cover a backyard greenhouse in one piece, practically eliminating all seams, and you may be able to cover the greenhouse several times from one roll.

Film-covered greenhouses do not need overly strong framing because the dead-weight of the film is very light. If the greenhouse is properly designed, there's no snow load problem because the snow will slide off rather than pile up. And the structure can flex enough to survive violent winds (unlike a necessarily rigid glass-covered greenhouse).

Polyvinyl chloride film is also used for greenhouses to some extent. Used in 8 mil thickness, it lasts up to five years but costs more than polyethylene. It comes only in widths up to 6 feet, requiring seams and overlaps in covering. Vinyl film, like a long-playing record, attracts dust and dirt from the air, and the cover will require cleaning periodically, not only for appearance, but also to maintain light-transmission efficiency. Polyester film (Mylar) is usable as a greenhouse cover, with a life of up to five years, but it is expensive.

FRAMING

Greenhouses with glass and rigid plastic panels are framed in wood, aluminum, or a combination of steel pipe and wood. Pipe-and-wood frames are used mainly for large commercial greenhouses. Film-covered greenhouses are framed with wood, aluminum, rigid plastic, and steel electrical conduit.

Plant growth is directly related to the amount of light received.

COVER MATERIALS FOR GREENHOUSES

Cover Material	Greenhouse Cover Thickness	Weight, lbs./sq. ft.	Clarity	Cost/sq. ft.	Normal Life
Films					
Polyfilm (Polyethylene)*	6 mil	0.03	milky translucent	$0.02	3 months
Ultraviolet Inhibited Polyethylene	4 mil to 6 mil	0.02 to 0.03	milky translucent	0.03 to 0.04	1 year
Vinyl (Polyvinyl-chloride, PVC)	8 mil to 16 mil	0.05 to 0.08	transparent	0.06 to 0.09	3 years
Polyester (Mylar)	7 mil	0.05	transparent	0.18	5 years
Rigid Plastics					
Fiberglass (polyester resin containing glass fibers)	.04 to .06	0.3 to 0.5	translucent	0.40 to 0.45	10–20 years
Acrylic (Plexiglas)	.08 to .188 inch	0.5 to 1.2	transparent	0.95 to 1.65	20 years
Polycarbonate (Lexan)	.08 to .125 inch	0.5 to 0.8	transparent	2.10 to 2.75	5–7 years
Glass					
Single Strength (SSB)	$\frac{3}{32}$ inch	1.2	transparent	0.30	indefinite
Double Strength (DSB)	$\frac{1}{8}$ inch	1.7	transparent	0.42	indefinite
Wired	$\frac{1}{4}$ inch	3.2	transparent	4.00	indefinite
Tempered	$\frac{1}{8}$ inch	1.6	transparent	6.50	indefinite

*Not recommended for greenhouse use.

This is why greenhouse roofs are designed to admit as much light as possible in relation to the load they must carry. This load includes the weight of the glazing material, the weight of the framing itself, snow and ice loads, and wind loading. Aluminum extrusion framing is less bulky than wood, but it costs more initially and results in greater heat loss than through a wood frame. Arguments that aluminum requires less maintenance than wood do not apply if the wood is construction-heart redwood. While other woods require periodic painting to preserve them, redwood does not. There is no paint or varnish you can apply that is more resistant to weathering and moisture than the bare redwood itself.

The frame of a film-covered greenhouse is accessible for periodic painting and repair every time the old cover is removed and replaced.

GREENHOUSE SIZE

In deciding how big a greenhouse you should buy or build, there are several factors to consider. First, many people wish they had bought a bigger greenhouse after a year or so of operation. Although some greenhouse kits are sold with the claim that you can add sections, enlarging the foundation after the greenhouse is up may not be simple and easy.

This gazebo greenhouse comes in six-foot hexagonal and eight-foot octagonal kits, with glass sides and fiberglass roof. The kits come preassembled in large sections. (*Courtesy Sturdi-built Manufacturing Co.*)

Second, if you operate the greenhouse twelve months a year, in most parts of the country you will have to heat it on winter nights and on some days, too. Winter heating is the largest part of greenhouse operating costs.

Third, how much space do you really have for a greenhouse? And where is it? The initial step is to work out possible locations for a greenhouse on your property. You should take into account sunlight, wind protection, zoning requirements. Also determine the maximum size greenhouse you can erect at the site. You can estimate the cost of heating a future greenhouse by interpolating from kit manufacturers' data or by your own calculations in accordance with those explained in Chapter 6. Then, you'll probably be happiest in the long run if you go to the biggest greenhouse you can afford to erect and operate.

SITE

Choosing a site for your greenhouse is not a simple matter of picking a spot with an unobstructed southern exposure for maximum winter sunlight. There are other factors. However, winter sunlight is the prime consideration. This sunlight is not only necessary for plant growth; it is also the major source of winter heat. Southern exposure is best, followed by southeastern, southwestern, eastern, and western in that order. A northerly exposure can be used, but that limits the range of plants that can be grown successfully. A minimum of four hours of direct sunlight is usually recommended, but within limits, growth will be proportional to hours of sunlight. Also the more hours of sunlight on the greenhouse, the fewer hours you will be running your heating system. When the sun is shining, it will supply all the heat needed even on the coldest winter day, and much of the time the heat may require that you open vents. Thus, solar heating for a greenhouse is certainly feasible.

Wind-chill factors apply to greenhouses too. A greenhouse exposed to the wind will cost more to heat than one that isn't. Siting the greenhouse behind a hedge or fence or in the lee of your house can materially reduce heating costs. Privet and juniper hedges make good windbreaks.

For even distribution of light, gabled greenhouses should be sited with the ridge running north and south.

In the summer, the greenhouse will require shading to control the heat buildup inside. This shading can be accomplished several ways,

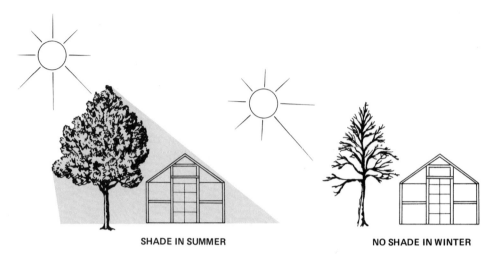

SHADE IN SUMMER   NO SHADE IN WINTER

A tree's shade can provide summer temperature control. In winter, the bare branches allow sunlight to enter, and the tree can diffuse winds somewhat.

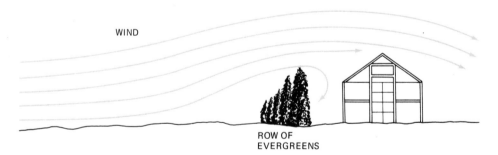

WIND

ROW OF
EVERGREENS

A windbreak consisting of a fence or row of small evergreens can provide protection from the prevailing winter wind and reduce heating costs.

but the easiest is to locate the greenhouse under deciduous trees.

The greenhouse site should have good drainage, and not be in the path of surface water draining from higher ground. Excess water from plant watering must not be allowed to collect on the floor. If the greenhouse has to be located on sloping ground, it may be necessary to raise the floor inside a masonry foundation to divert run-off water.

On the other hand, greenhouses built with floors below ground level—pit greenhouses—have advantages too. They cost less to heat, and cost less to build. But good drainage is essential.

BUILDING PERMITS

If you live in a locality where construction is regulated by building code and zoning restrictions, it is likely that you will need a construction permit. In some localities, yard structures such as tool sheds and greenhouses are exempt from code requirements if they are smaller than some specified size (typically 100 square feet) and are not anchored in concrete or masonry. This exception notwithstanding, you may well need a construction permit even for a window greenhouse.

Before you lay out any money for a kit greenhouse or begin any construction, get a copy of the building code and a copy of the zoning code for your area, and read them carefully. The purpose of the building code is to set minimum construction standards to protect the health and safety of the building's occupants and the community in general. The zoning code sets land use standards. Whether the codes are state, county, town, or village, enforcement will be up to a local building inspector, if there is one. If there is no local building inspector, there is unlikely to be any enforcement, but for good measure and to enhance the resale value of your property, build in accordance with the code.

Few building codes have specific construction requirements for backyard greenhouses. Most greenhouses are classified as miscellaneous accessory buildings usual to a residence, the same classification that a garden tool shed would fall under.

Whether you propose to erect a kit greenhouse or one of your own design, your foundation will have to meet the building code requirement for materials, footing, and depth, regardless of what the greenhouse manufacturer has to say about the foundation required under his product. The local code is based on area soil conditions and frost line;

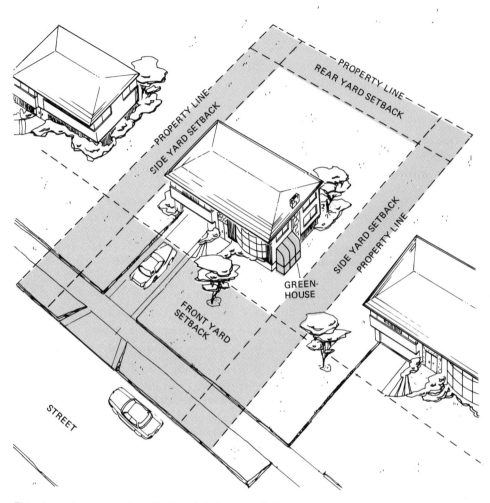

This shows how property setbacks relate to property lines.

it would be good common sense, if nothing else, to observe its requirements. If you are going to pump water to the greenhouse, or run a branch electric circuit, the building code will contain requirements.

Placement of the greenhouse on your property will have to meet front, side, or backyard setback requirements. Setbacks restrict how close you can build to the edge of your property. If where you want to put the greenhouse violates setback requirements, then you will have to seek a variance, which involves hearings, testimony from your immediate neighbors that the variance is fine by them, and often additional requirements that may include hassle and expense.

SAMPLE APPLICATION
FOR A BUILDING PERMIT

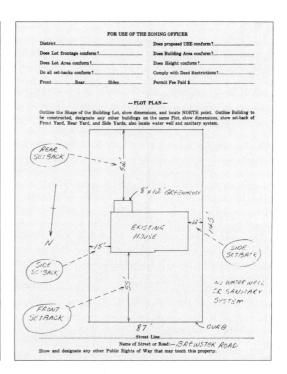

Local codes will probably require that you obtain a building permit before erecting your greenhouse. At left, is the front side of a typical application form. At right, is the form's backside, with plot plan. You may also have to prepare a simple drawing showing construction details.

Your first step, after studying the codes, should be to draw a sketch of what you think you want to do and present it to the local building official for an informal discussion. You could get some helpful advice. You can then obtain an application for a building permit. The application asks what is already on your property, what you plan to build, and when you will start and finish construction. You will also be asked to draw a plot plan, which is a dimensioned sketch of your property showing the location of all buildings and setbacks, and showing where you desire to put the greenhouse. You will be asked to estimate the cost of construction, and provide the building officials with a set of plans showing in detail what you plan to build. If you are erecting a kit greenhouse, you will probably only have to show details of the foundation, and wiring and plumbing if included. If everything is in order, you will be issued a building permit for which you will pay a fee based on the cost of construction. When the greenhouse is finished you then apply for a certificate of occupancy and use, the fee for which is usually included in the building permit fee. There are usually requirements to inform the officials when you are ready to have certain stages of construction inspected—such as footings, foundations prior to back filling, and completion. Once the certificate of occupancy is issued, you can expect your property tax to go up. After all, the value of your property has just been increased by one greenhouse!

OPERATING A GREENHOUSE

The essential purpose of a greenhouse is to provide a totally controlled environment for your plants. This is what separates a greenhouse gardener from an outdoors gardener. An outdoors gardener can blame everything under the sun for a crop failure. As a greenhouse gardener, you control the environment of your plants. You, not nature, are in charge.

The shell of a greenhouse gives your plants physical protection from the elements and allows plenty of sunlight to reach them, but before you start moving them in, you have some additional things to take care of. You must also control temperature, humidity and ventilation. (You must also control pests and diseases, but these are not primarily problems of erecting and equipping a greenhouse. So we'll not go into them here.)

While solar energy is the principal source of heat for your greenhouse, in cold weather an artificial source of heat must be provided to keep the greenhouse at some reasonable growing temperature at night, and on cloudy days. Not all plants require the same growing temperature. Ventilation is necessary to reduce heat buildup during sunlit hours, and to replenish the carbon dioxide in the greenhouse air. Natural ventilation often must be augmented by forced ventilation from either exhaust or inlet fans.

Greenhouse daytime temperatures also can be controlled by shading. While commercial greenhousemen often apply cheap shading

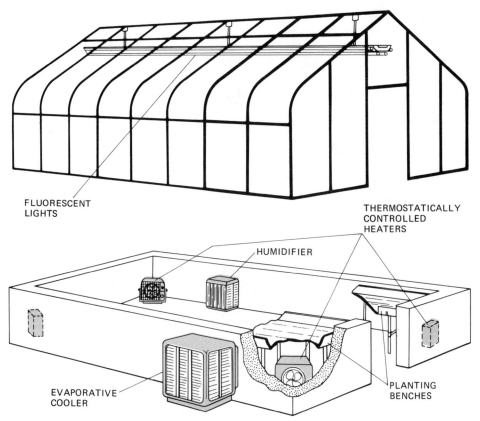

FLUORESCENT
LIGHTS

THERMOSTATICALLY
CONTROLLED
HEATERS

HUMIDIFIER

EVAPORATIVE
COOLER

PLANTING
BENCHES

Greenhouse kits are usually sold as shells with only a frame and cover. Or they are often sold with a minimum of equipment. To the shell you may want to add equipment to control temperature, humidity, ventilation, and shade. Amenities such as benches, work tables, running water, and lighting may be needed too.

paint to the glass, this produces an unsightly mess and can't be applied to plastic covers anyway. Shading in the home greenhouse is usually provided by trees. While trees are natural and may be placed well at the site, their shade cannot be controlled on a day-to-day basis. One other means of shade control includes roll-up or roll-down shades of aluminum or wood slats. Or you can use saran plastic or vinyl shading film that is cut to fit the glass lights and squeegeed to the inside surface.

While ventilation can reduce the inside temperature of a greenhouse, it can never get the temperature quite all the way down to the

This small window-mounted green-house jutting over narrow alley gets light from its sides and top. (*Adolf deRoy Mark, Architect*)

outside temperature. And in the summer the outside temperature may be too high for some of your plants. Wet-pad, or evaporative-cooling, equipment can lower the greenhouse temperature close to the outside shade temperature.

In the winter, just like in your house, the humidity in a sun-warmed or artificially-heated greenhouse tends to get too low for most plants. This can be corrected by extra watering or by wetting down the greenhouse floor and walkways several times during the day, or you can install a humidifier.

Again, carbon dioxide depletion is one reason for ventilation. Sometimes on winter days, you should not open the vents because it is simply too cold and windy. But carbon dioxide can be easily added to the greenhouse air by use of dry ice or compressed gas, or by burning alcohol.

You can control your greenhouse environment by hand, running around reading thermometers and other instruments, opening and closing vents, and adjusting heaters, humidifiers and other equipment. Or you can automate some or all of the controls. Greenhouse climate control equipment is discussed in detail in the next chapter.

22

This greenhouse enabled the owners to get some light into a dark cellar. The narrow courtyard, shown in the drawing, was excavated to half the depth of the cellar floor. The greenhouse floor forms a well-lit landing halfway between the first floor and the finished cellar. (*Adolf deRoy Mark, Architect*)

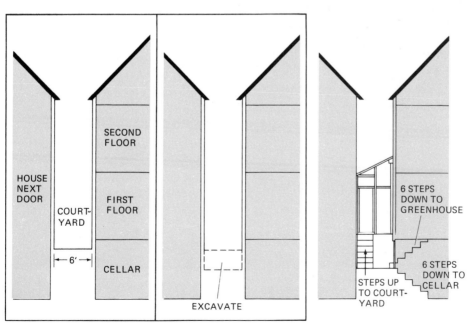

HOUSE NEXT DOOR

COURT-YARD

SECOND FLOOR

FIRST FLOOR

CELLAR

← 6' →

EXCAVATE

6 STEPS DOWN TO GREENHOUSE

STEPS UP TO COURT-YARD

6 STEPS DOWN TO CELLAR

Kit greenhouses are adaptable to unusual installations. Shown is a Lord & Burnham Orlyt Imperial even-span model in a location where it at first appeared that only a lean-to design could be used.

KITS OR DO-IT-YOURSELF DESIGNING

Even with all the greenhouse kits on the market, there are several reasons why you might choose not to use one. A kit greenhouse in the style you like may not be suitable for the site you want to use, such as in a corner of your patio or on top of your garage roof. You can save money by designing and building your own, as compared to the cost of a kit, but some know-how is required.

USING AN ARCHITECT

Greenhouses that become a part of a major house remodeling offer special challenges. Unless you are particularly knowledgeable about remodeling and structural changes, you'd be wise to consult an architect. This is not an occasion to call in a carpenter or home building contractor for advice or to let him do the job. Few of these craftsmen have

This architect-designed greenhouse is built over a garage and opens into the end of a second-floor living room. (*Adolf deRoy Mark, Architect*)

Since lean-to greenhouses save yard space, they are especially popular in urban settings. Here is an attractive design for tight quarters. (*Adolf deRoy Mark, Architect*)

the training or experience for such a task. A good architect can create plans for a good looking, yet functional, greenhouse.

An architect's help can be valuable. The title "architect" is legal, just like "doctor" or "dentist." An architect has had training and has demonstrated his professional competence by examination. He is licensed or registered by the state to practice architecture. An architect can do the following for you:

- Draw up preliminary sketches showing the general plan and appearance of your project and then make modifications you want.
- Prepare preliminary cost estimates.
- Prepare working and detail drawings and specifications from which you can build the greenhouse yourself. Or
- Advise you on the selection of contractors or subcontractors, and analyze competitive bids.

● Supervise construction and evaluate a contractor's suggested changes in design and materials.
● Advise you whether a contractor has followed plans and specifications, and thereby earned partial or full payment.

Don't hesitate to ask an architect about details of his services. The fee for an architect's performing the full range of services is usually about 10 to 20 percent of construction, depending on the size and complexity. When only some of the services are performed, as when you do the construction yourself, the fee should be less. An architect's services may also be obtained on an hourly or negotiated basis. Usually, a retainer is paid when the contract is signed. The rest is paid as contracted.

A large patio facing south, southwest, or southeast is an ideal site for a lean-to greenhouse. Shown is the Janco Highlander lean-to in a glass-to-ground version. Such greenhouses provide more growing space than those built atop masonry walls, and they cost about the same to heat. The 10 percent higher initial cost of the glass-to-ground style is more than offset by a lower total masonry cost. (*Courtesy J. A. Nearing Co., Inc.*)

2

Greenhouses
You Can Build from Kits

Kit greenhouses range from tiny film-covered huts to large, elegant houses with curved-glass eaves. In between these extremes are medium-sized models covered with film, rigid plastic, or glass. Styles include even-span gables, lean-tos, A-frames, and domes. Unless you have a difficult space problem, or want a greenhouse with exotic style or construction, there is probably a kit greenhouse to match your requirements.

For anyone with a reasonable amount of home-repair or do-it-yourself experience, assembling any of the kit greenhouses should present no big problem. Structural parts, whether wood or metal, will be cut to dimension. But you will have to prepare your own greenhouse foundation (see Chapter 4). Most kits are shells only. For these you must provide all of the climate-control equipment as well.

If you are planning to buy a kit greenhouse, your first step should be to study catalogs carefully. There are great differences between makes of kits. After you have narrowed the possibilities down, see if you can arrange to see a sample of the greenhouses you are considering. Ask the manufacturer who has one in your area. You can locate addresses of greenhouse manufacturers in the appendix of this book.

The greenhouse kits described in the following pages are a representative sampling of what is available. And the variety of kits for greenhouses covered with glass is far greater than that for greenhouses covered with rigid plastic.

The distinctive style of the Sunflare kit greenhouse is suited to a wide variety of home architecture. The greenhouse is 15 feet in diameter. The kit comes with base wall, greenhouse, glass, cantilevered benches, and automatic hydraulic dome ventilation. (*Courtesy Sturdi-built Manufacturing Co.*)

The Eaglet free-standing gable greenhouse is typical of medium-size glass and aluminum English-style greenhouses available as kits from many sources. (*Courtesy National Greenhouse Co.*)

The Sun Dome II kit comes with triangles factory-assembled and covered. This geodesic dome is 15 feet in diameter. No foundation is required. (*Courtesy Peter Reimuller—The Greenhouseman*)

FILM-COVERED GREENHOUSES

Greenhouses covered with polyethylene film are very practical for either the beginner or the experienced home gardener. They are made in free-standing and lean-to designs with floor areas ranging from 75 square feet to 476 square feet. The structure under the cover can be temporary or permanent, but the polyethylene cover must be replaced, usually every year. This can be a nuisance. Although film-covered greenhouses cannot compete with glass houses for all-around attractiveness, they are an economical approach.

Vinyl film is also used on hobby greenhouses, particularly in geodesic domes. It is crystal-clear, giving the appearance of glass, but it must be replaced every few years.

National Econolite

This is a Quonset-shaped model that is framed with arched ¾-inch electrical conduit made of galvanized steel. Two layers of 4-mil polyethylene cut down heat loss and reduce inside condensation. Gable ends are braced with 2 × 4 redwood (supplied). Arches are staked into the ground.

KITS FOR POLYETHYLENE FILM-COVERED GREENHOUSES

Brand and Model	Style	Length	Width	Height Maximum	Height Eave
National Econolite	Quonset	10' 0"	12' 0"	7' 0"	
	Quonset	15' 0"	12' 0"	7' 0"	
	Quonset	20' 0"	12' 0"	7' 0"	
Peter Reimuller					
Crystalaire	Gothic	8' 0"	6' 6"		
Crystalaire	Gothic	12' 0"	6' 6"		
Sun Dome I	Dome	10' 0" diameter			
Turner Greenhouses					
Model 1414P	Gable	14' 0"	14' 0"	8' 0"	5' 10"
Model 1418P	Gable	18' 0"	14' 0"	8' 0"	5' 10"
Model 1422P	Gable	22' 0"	14' 0"	8' 0"	5' 10"
Model 1426P	Gable	26' 0"	14' 0"	8' 0"	5' 10"
Model 1430P	Gable	30' 0"	14' 0"	8' 0"	5' 10"
Model 1434P	Gable	34' 0"	14' 0"	8' 0"	5' 10"
Model 714P	Lean-to	14' 0"	7' 0"	8' 0"	5' 10"
Model 718P	Lean-to	18' 0"	7' 0"	8' 0"	5' 10"
Model 722P	Lean-to	22' 0"	7' 0"	8' 0"	5' 10"
Model 726P	Lean-to	26' 0"	7' 0"	8' 0"	5' 10"
Model 730P	Lean-to	30' 0"	7' 0"	8' 0"	5' 10"

Note: 1. Prices include a blower for inflating the cover.

Peter Reimuller Crystalaire

This unit has a gothic arch formed by bent hardboard. The rest of the frame is redwood. All wood comes precut and drilled for bolt fastening. The Dutch door has an opening in the bottom half for the installation of optional 10-inch square automatic inlet shutters. The framing in back wall permits mounting a 10-inch exhaust fan.

Floor Area Square Foot	Price	Cost per Square Foot	Film Thickness	Frame Material	Note
120	$368	$3.06	2 layers 4 mil	Steel	1
180	408	2.27	2 layers 4 mil	Steel	1
240	463	1.93	2 layers 4 mil	Steel	1
52	125	2.40	8 mil	Redwood & Hardboard	
78	145	1.86	8 mil	Redwood & Hardboard	
79	135	1.70	8 mil	Redwood	
196	249	1.27	6 mil	Steel	
252	303	1.20	6 mil	Steel	
308	357	1.16	6 mil	Steel	
364	411	1.13	6 mil	Steel	
420	465	1.11	6 mil	Steel	
476	519	1.09	6 mil	Steel	
98	144	1.47	6 mil	Steel	
126	171	1.36	6 mil	Steel	
154	198	1.29	6 mil	Steel	
182	225	1.24	6 mil	Steel	
210	252	1.20	6 mil	Steel	

This film-covered Crystalaire gothic greenhouse has redwood framing with heat-bonded hardboard arches and an 8-mil polyethylene cover. Assembly is with bolts. No foundation is needed. (*Courtesy Peter Reimuller—The Greenhouseman*)

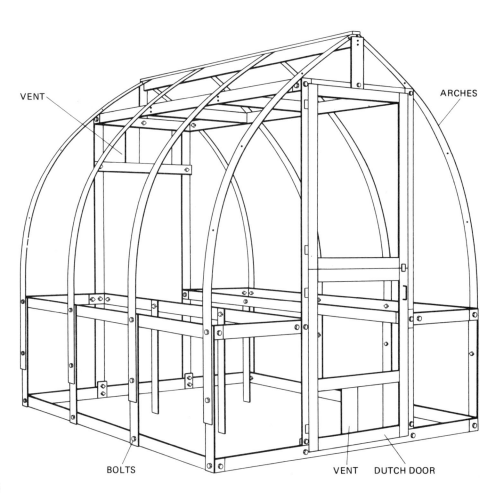

VENT

ARCHES

BOLTS

VENT   DUTCH DOOR

Peter Reimuller Sun Dome I

This has a five-side dome constructed of fifteen triangles with double-strut construction which allows sides of the triangles to be bolted together, clamping the polyethylene film. Ventilation is provided by a vent in the top and a sliding vent in the door. Frame parts are cut to dimension and drilled. The door is prefabricated and has an integral sill.

Turner Greenhouses

The framework is made of steel, coated with a rust-resistant aluminum-zinc alloy. The polyethylene cover is mounted in three pieces, one for the top and sides, and one for each end, by means of rigid hold-down caps, aluminum fasteners, and pressure-sensitive tape. These greenhouses can be outfitted with fiberglass covers by means of a conversion kit which consists of cover, bars and fasteners. There is a manually operated vent in each end. The door is polyethylene-covered metal. Automatic exhaust fans for each end are optional. The manufacturer recommends using a wood sill (not supplied) or concrete block foundation.

FIBERGLASS-COVERED GREENHOUSES

Here the cover material is actually a polyester-resin strengthened with glass fibers and fortified with acrylic. For some brands and formulations, manufacturers claim a 15- to 20-year life expectancy. The translucence of fiberglass covers used varies widely. If you are buying a mail-order kit, you'd be wise to request a sample of the cover before placing your order.

McGregor

Redwood frame parts are precut but not drilled. Assembly is with steel plates and bolts. Footings are not required. Anchors for burial in earth are provided.

Corrugations in the cover run from ridge to cove on the roof and top to bottom on the sides; this minimizes dirt, leaf, snow, and ice accumulation. Corrugations add strength. This cover is guaranteed for 12 years.

The Oasis greenhouse has a corrugated fiberglass cover. The unit is modeled after a Virginia Polytechnic Institute, do-it-yourself design shown in the next chapter. (*Courtesy Texas Greenhouse Co., Inc.*)

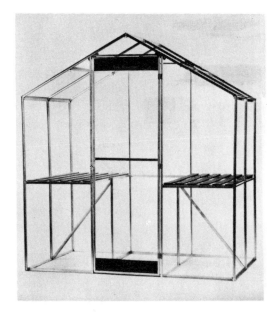

This simple extruded-aluminum frame for Sun/America's Model 1100 "starter" has a single-bolt assembly at each joint. The frame can be covered with clear vinyl, which will outlast polyethylene film by many seasons. (*Courtesy Sun/America Corp.*)

This Plantsman free-standing greenhouse has a corrugated fiberglass cover and a redwood frame. It is available in several sizes. The advantages of straight sides vs. sloping sides are debatable. (*Courtesy McGregor Greenhouses*)

This Pearl Mist lean-to is covered with corrugated fiberglass. It has a redwood frame, a Dutch door, and is made in three sizes. The smallest size is shown here. Free-standing models are also available. (*Courtesy Peter Reimuller—The Greenhouseman*)

KITS FOR VINYL-COVERED GREENHOUSES

Brand and Model	Style	Length	Width	Height Maximum	Height Eave
Casaplanta					
Model CP-1	Gothic	4' 0"	6' 0"	7' 0"	—
Sears, Roebuck					—
Model 32KF4960N	Gable	3' 0"	8' 0"	7' 0"	—
Model 32KF49622N2	Gable	6' 0"	8' 0"	7' 0"	—
Model 32KF49623N3	Gable	9' 0"	8' 0"	7' 0"	—
Model 32KF49624N4	Gable	12' 0"	8' 0"	7' 0"	
Sun/American					
Model 1100	Gable	4' 0"	6' 0"	6' 8"	5' 0"
Model 1120	Gable	8' 0"	6' 0"	6' 8"	5' 0"
Model 1140	Lean-to	4' 0"	4' 0"	7' 2"	5' 0"

Peter Reimuller

Redwood frames are precut and drilled. The fiberglass is attached to frames with screw nails. The Dutch door frame is preassembled. Corrugations in the cover run parallel to the roof ridge, and run top to bottom on the sides. Advertised life of the cover is 10 to 15 years.

Texas Greenhouse

This Virginia Polytechnic Institute design has curved rafters of laminated redwood. Cover choice includes sky green fiberglass (81 per cent light transmission) or clear translucent (82 percent light transmission). There is an aluminum combination storm door as well as provision for mounting a thermostatically-controlled 12-inch shuttered exhaust fan on the rear gable and motor-driven inlet shutters on the door-end gable. Accessories include benches and saran shade fabric.

Turner Greenhouse

There are manually-operated vents in each end. And automatic exhaust fans are available. The fiberglass-covered door has a metal frame. The manufacturer recommends a wood sill (not supplied) or a concrete block foundation.

Floor Area Square Foot	Price	Cost per Square Foot	Film Thickness	Frame Material	Note
24	$ 75	$ 3.13	8 mil	Extruded plastic tubing	Redwood benches included
24	250	10.41	16 mil	Aluminum	
48	350	7.29	16 mil	Aluminum	
72	530	7.36	16 mil	Aluminum	
96	670	6.98	16 mil	Aluminum	
24	150	6.25	12 mil	Aluminum	
48	250	5.21	12 mil	Aluminum	
16	110	6.88	12 mil	Aluminum	

The Standard Greendome has an aluminum framework, power ventilation, a thermostat, automatic louvers, and an interior light. The cover is clear vinyl or vinyl on Dacron membrane. Cost for a 20-foot dome is $2,800. A second-skin option for insulation adds $800. (*Courtesy Dome East Corp.*)

RIGID PLASTIC-COVERED GREENHOUSES

Rigid, flat plastic panels are used in a variety of ways in greenhouse covers besides being substituted one-for-one for glass lights. Expected life, appearance, and light transmission vary.

Dome East

Greendome is an adaptation of the company's line of geodesic shelterdomes. The frame is aluminum tubing joined with hub plates. The vinyl or vinyl and dacron cover is suspended below the frame. A

KITS FOR FIBERGLASS-COVERED GREENHOUSES

Brand and Model	Style	Length	Width	Height Maximum	Height Eave
McGregor					
Plantsman No. 4	Gable	4' 2"	6' 10"	8' 0"	5' 9"
Plantsman No. 8	Gable	8' 4"	6' 10"	8' 0"	5' 9"
Plantsman No. 12	Gable	12' 3"	6' 10"	8' 0"	5' 9"
Plantsman No. 16	Gable	16' 5"	6' 10"	8' 0"	5' 9"
Peter Reimuller					
Pearl Mist 8 × 8	Gable	8' 0"	8' 0"	7' 6"	6' 8"
Pearl Mist 8 × 12	Gable	12' 0"	8' 0"	7' 6"	6' 8"
Pearl Mist 8 × 16	Gable	16' 0"	8' 0"	7' 6"	6' 8"
Pearl Mist "Eight"	Lean-to	8' 0"	7' 0"	7' 8"	6' 5"
Pearl Mist "Twelve"	Lean-to	12' 0"	7' 0"	7' 8"	6' 5"
Pearl Mist "Sixteen"	Lean-to	16' 0"	7' 0"	7' 8"	6' 5"
Texas					
Oasis	Gothic	11' 0"	10' 0"	8' 5"	—
Turner					
Model 1414 FG	Gable	14' 0"	14' 0"	8' 0"	5' 10"
Model 1418 FG	Gable	18' 0"	14' 0"	8' 0"	5' 10"
Model 1422 FG	Gable	22' 0"	14' 0"	8' 0"	5' 10"
Model 1426 FG	Gable	26' 0"	14' 0"	8' 0"	5' 10"
Model 1430 FG	Gable	30' 0"	14' 0"	8' 0"	5' 10"
Model 1434 FG	Gable	34' 0"	14' 0"	8' 0"	5' 10"
Model 714 FG	Lean-to	14' 0"	7' 0"	8' 0"	5' 10"
Model 718 FG	Lean-to	18' 0"	7' 0"	8' 0"	5' 10"
Model 722 FG	Lean-to	22' 0"	7' 0"	8' 0"	5' 10"
Model 726 FG	Lean-to	26' 0"	7' 0"	8' 0"	5' 10"
Model 730 FG	Lean-to	30' 0"	7' 0"	8' 0"	5' 10"

second cover can be suspended inside the cover for improved thermal insulation. Greendome is supplied with a power ventilation system (no roof vents are possible), an aluminum doorway, an automatic thermostat, an interior light, automatic louvers, anchoring hardware, and a shade panel.

Grow House (Sears, Roebuck & Co.)

The cover is double-layer plastic board consisting of two thin walls separated by close-spaced parallel ribs. This material resembles corrugated cardboard, the ribs creating a dead air space that provides

Floor Area Square Foot	Price	Cost per Square Foot	Type of Fiberglass	Frame Material
28	$ 185	$6.61	4 oz. Filon	Redwood
57	240	4.21	4 oz. Filon	Redwood
84	325	3.87	4 oz. Filon	Redwood
112	400	3.57	4 oz. Filon	Redwood
64	245	3.83	Glasteel	Redwood
96	335	3.49	Glasteel	Redwood
128	415	3.24	Glasteel	Redwood
56	215	3.84	Glasteel	Redwood
84	305	3.63	Glasteel	Redwood
112	385	3.44	Glasteel	Redwood
110	595	5.40	5 oz. Fiber-glass	Redwood
196	667	3.40	Not Specified	Steel
252	803	3.19	Not Specified	Steel
308	939	3.05	Not Specified	Steel
364	1,075	2.95	Not Specified	Steel
420	1,211	2.08	Not Specified	Steel
476	1,483	2.23	Not Specified	Steel
98	350	3.57	Not Specified	Steel
126	418	3.32	Not Specified	Steel
154	486	3.16	Not Specified	Steel
182	554	3.04	Not Specified	Steel
210	622	2.96	Not Specified	Steel

This lean-to greenhouse has a two-layer polyboard cover which diffuses sunlight evenly and reduces winter heat loss. (*Courtesy Grow House Corporation*)

good thermal insulation. The frame is aluminum. Accessories include manually operated screened vents, a 12-inch fan-operated vent kit, and shelves.

Peter Reimuller

Triangles of geodesic dome are factory assembled with a fiberglass/acrylic cover. Site assembly consists of bolting triangles together onto a wooden sill. All framing is redwood. A large vent in the top provides convection ventilation when the door is also open. Accessories include an electric heater and powered ventilation.

Sturdi-built

Frames are redwood, and covers are either of clear or shading fiberglass. Unlike other kit greenhouses which are supplied in individual parts, these are prefabricated in large sections. They are available in a variety of lengths in two-foot increments. Perimeter masonry foundation is required. Roof vents provide convection cooling.

Sun-Bon greenhouse by Sturdi-built shows that a greenhouse does not have to have a plain, strictly functional appearance to be effective in growing plants. This greenhouse has a redwood frame and a choice of clear or self-shading fiberglass. It can be dismantled and moved. (*Courtesy Sturdi-built Manufacturing Co.*)

Vegetable Factory's lean-to fits a deck area as if custom designed, yet it is a standard model. The greenhouse has a two-layer flat fiberglass cover bonded in panels to aluminum extrusion. (*Courtesy Vegetable Factory*)

FLAT RIGID-PLASTIC-COVERED GREENHOUSE

Brand and Name	Style	Length	Width	Height Maximum	Height Eave
Dome East					
Greendome	Dome	20' diameter		7' 6"	—
Grow House (Sears, Roebuck & Co.)					
Model 4955IN	Gable	5' 0"	7' 0"	6' 10"	4' 7"
4955IN + 49552N	Gable	10' 0"	7' 0"	6' 10"	4' 7"
49554N	Lean-to	6' 0"	4' 0"	7' 0"	4' 7"
Sun/America					
Model 1200	Gable	5' 0"	7' 0"	6' 10"	
Model 2300	Lean-to	6' 0"	4' 0"	7' 0"	
Peter Reimuller					
Star Dome II	Dome	15' diameter		—	—
Sturdi-built					
Sun-bon	Gable	10' 0"	8' 0"	8' 0"	7' 0"
Sun-bon	Gable	10' 0"	9' 0"	8' 0"	7' 0"
Sun-bon	Gable	10' 0"	10' 0"	8' 0"	7' 0"
Filteray**	Gable	12' 0"	11' 0"	8' 0"	6' 0"
Filteray**	Gable	20' 0"	11' 0"	8' 0"	6' 0"
Filteray**	Lean-to	12' 0"	7' 6"	8' 0"	6' 0"
Filteray**	Lean-to	20' 0"	7' 6"	8' 0"	6' 0"
Filteray**	Lean-to	12' 0"	9' 6"	9' 0"	6' 0"
Filteray**	Lean-to	20' 0"	9' 6"	9' 0"	6' 0"
Vegetable Factory					
Model 8008	Gable	8' 1"	8' 2"	7' 4"	6' 0"
Model 8012	Gable	12' 0"	8' 2"	7' 4"	6' 0"
Model 1108	Gable	8' 1"	11' 0"	8' 1"	6' 0"
Model 1112	Gable	12' 0"	11' 0"	8' 1"	6' 0"
Model 5508	Lean-to	8' 1"	5' 6"	8' 1"	6' 0"
Model 5512	Lean-to	12' 0"	5' 6"	8' 1"	6' 0"
Model 8500	Lean-to	8' 1"	11' 8"	9' 2"	6' 0"
Model 8512	Lean-to	12' 0"	11' 8"	9' 2"	6' 0"
Model 1712	Gable	12' 0"	17' 3"	9' 2"	6' 0"

Notes: *Price includes power ventilation, shade panel. Second skin option costs $3,870.

      **Intermediate sizes available in 2-foot length increments.

Floor Area Square Foot	Price	Cost per Square Foot	Cover Material	Frame Material
300	$7,800*	$26.00	Vinyl	Aluminum
35	235	6.71	Plastic Board	Aluminum
70	420	6.00	Plastic Board	Aluminum
24	185	7.71	Plastic Board	Aluminum
35	250	7.14	Double-wall Polyboard	Aluminum
24	200	8.30	Double-wall Polyboard	Aluminum
177	650	3.67	Acrylic/Fiberglass	Redwood
80	1,320	16.50	Fiberglass	Redwood
90	1,430	15.89	Fiberglass	Redwood
100	1,610	16.10	Fiberglass	Redwood
132	1,260	9.55	Fiberglass	Redwood
220	1,660	7.55	Fiberglass	Redwood
90	910	10.11	Fiberglass	Redwood
150	1,160	7.73	Fiberglass	Redwood
114	1,135	9.96	Fiberglass	Redwood
190	1,395	7.34	Fiberglass	Redwood
66	1,055	15.98	Acrylic/Fiberglass	Aluminum
98	1,359	13.87	Acrylic/Fiberglass	Aluminum
89	1,550	17.42	Acrylic/Fiberglass	Aluminum
132	1,945	14.73	Acrylic/Fiberglass	Aluminum
44	699	15.89	Acrylic/Fiberglass	Aluminum
66	895	13.56	Acrylic/Fiberglass	Aluminum
94	1,175	12.50	Acrylic/Fiberglass	Aluminum
140	1,459	10.42	Acrylic/Fiberglass	Aluminum
207	2,850	13.77	Acrylic/Fiberglass	Aluminum

Vegetable Factory

The cover is a half-inch thick panel assembly consisting of an outer and inner layer of fiberglass bonded in panels to H-shaped aluminum extrusion frames. These panels are then mounted in the aluminum frame of the greenhouse. Then the greenhouse may be anchored to any flat surface. Ventilation can be provided by optional inlet louvers and an accessory exhaust fan.

ENGLISH STYLE GREENHOUSES

In England, where gardening under glass is enjoyed by millions, more people own this style of greenhouse than any other. They are the smallest type of glass-covered greenhouses you can buy except for window models. They are made in both gable (even span) and lean-to types and all have glass-to-ground vertical sides. Frames are normally aluminum, but one manufacturer has substituted a combination of redwood and aluminum. Framing members can be lighter than those used in large glass greenhouses because of shorter spans.

When free-standing, they can be erected on redwood sill plates anchored into the ground. Lean-tos and gables attached to a house should have masonry foundations going below the frost line to prevent heaving. Assembly is not difficult, but glass must be handled carefully. Doors and adjacent areas are glazed with tempered glass or acrylic for safety. Extensive selections of accessories available.

Burpee

This greenhouse line is imported from England by the W. Atlee Burpee Co. of Burpee seed fame. Double-strength glass is used except for safety glazing. Glass is held by clips, and the sliding door cannot slam in wind. Gable models have roof and side vents. Lean-tos have roof and end vents.

Lord & Burnham

The Sunlyt line of greenhouses has extruded aluminum frames and single-strength glass. Glass is held in by bar caps. Roof vents provide convection ventilation. Bench legs are standard equipment.

This imported English-style free-standing greenhouse has a full-height sliding door, an advantage in windy weather. A lean-to model is also available. (*Courtesy W. Atlee Burpee Co.*)

Small greenhouse lines of Texas Greenhouse combine redwood and aluminum framing. Glass is installed with glazing strips. (*Courtesy Texas Greenhouse Co., Inc.*)

National

Panalite greenhouse has an extruded aluminum frame, and double-strength glass held by bar caps. Standard equipment includes 2 × 4 sill plates, a jalousie window and screen, redwood benches, and redwood shelves.

ENGLISH STYLE GREENHOUSES

Brand and Model	Style	Length	Width	Height Maximum	Height Eave
Burpee					
8 × 12	Gable	12' 0"	8' 0"	7' 4"	4' 3"
8 × 8	Gable	8' 0"	8' 0"	7' 4"	4' 3"
6 × 12	Lean-to	12' 0"	6' 0"	7' 8"	4' 11"
6 × 8	Lean-to	8' 0"	6' 0"	7' 8"	4' 11"
Janco					
Belair 9	Gable	12' 9"	9' 6"	8' 2"	5' 6"
Belair 9	Gable	19' 0"	9' 6"	8' 2"	5' 6"
Belair 6	Gable	8' 8"	6' 8"	7' 5"	5' 6"
Belair 6	Gable	12' 9"	6' 8"	7' 5"	5' 6"
Belair 4	Lean-to	8' 8"	4' 5"	8' 0"	5' 6"
Belair 4	Lean-to	12' 9"	4' 5"	8' 0"	5' 6"
Lord & Burnham					
Sunlyt					
Even Span 9	Gable	12' 9"	9' 10"	8' 4"	5' 6"
Even Span 9	Gable	19' 0"	9' 10"	8' 4"	5' 6"
Even Span 6	Gable	8' 8"	6' 9"	7' 5"	5' 6"
Even Span 6	Gable	12' 9"	6' 9"	7' 5"	5' 6"
Lean-To 4	Lean-to	8' 8"	4' 7"	8' 2"	5' 6"
Lean-To 4	Lean-to	12' 9"	4' 7"	8' 2"	5' 6"
National					
Panalite	Gable	8' 6"	7' 6"	7' 11"	6' 2"
Panalite	Lean-to	8' 6"	5' 3"	8' 9"	6' 2"
Texas Greenhouse					
Little Gardener*					
Little Gardener L-5	Gable	10' 6"	8' 9"	8' 10"	6' 6"
Little Gardener L-6	Gable	12' 7"	8' 9"	8' 10"	6' 6"
Little Gardener L-13	Gable	27' 1"	8' 9"	8' 10"	6' 6"

*Available in approximately 2-foot increments.

Texas Greenhouse

The Little Gardener line of greenhouses has redwood frames. Ridge and eave joints are gusseted with aluminum plates. Double-strength glass is mounted in aluminum glazing strips. Multiple roof vents in all sizes provide natural ventilation with the vent in the door.

Square Feet	Price	Cost per Square Foot	Glass Thickness	Glass Retention	Frame
96	$ 895	$ 9.32	DS	Clips	Aluminum
64	675	10.54	DS	Clips	Aluminum
72	785	10.90	DS	Clips	Aluminum
48	645	13.43	DS	Clips	Aluminum
121	924	7.63	SS	Bar Caps	Aluminum
181	1175	6.49	SS	Bar Caps	Aluminum
58	620	10.67	SS	Bar Caps	Aluminum
85	795	9.35	SS	Bar Caps	Aluminum
38	495	13.02	SS	Bar Caps	Aluminum
56	585	10.45	SS	Bar Caps	Aluminum
125	1082	8.66	SS	Bar Caps	Aluminum
187	1338	7.16	SS	Bar Caps	Aluminum
59	710	12.03	SS	Bar Caps	Aluminum
86	882	10.26	SS	Bar Caps	Aluminum
40	548	13.70	SS	Bar Caps	Aluminum
58	648	11.17	SS	Bar Caps	Aluminum
64	630	9.84	DS	Bar Caps	Aluminum
45	490	10.89	DS	Bar Caps	Aluminum
92	744	8.09	DS	Clips	Redwood & Aluminum
110	833	7.57	DS	Clips	Redwood & Aluminum
237	1456	6.14	DS	Clips	Redwood & Aluminum

Janco

Belair greenhouses have an extruded aluminum frame. Single-strength glass is held by aluminum bar caps. Double-strength glass is optional. There are no roof vents. Jalousie windows in gable ends provide cross ventilation. Bench legs and side-support rails are standard equipment.

LARGE GLASS GREENHOUSES

This category includes what might be called "estate greenhouses," although you don't need an estate to use one to good advantage. Some people term them "professional greenhouses," because small commercial growers use them. But whatever you call them, they are big, top-of-the-line models that are expensive. When you buy one, you get a lot of greenhouse value for your money.

Large greenhouses are made in lean-to and free-standing even-span models in many widths, and all are available in lengths of set increments. Some are glass-to-ground and need only a shallow foundation. Others are designed to be erected on top of a low masonry perimeter wall. Their structural designs, and their availability in many sizes make them particularly adaptable for custom installation and unusual sites. Unlike smaller kits, which are essentially packaged before you

This window greenhouse is composed of two units designed to fit an extra-wide window. The units can be used separately too. (*Courtesy J. A. Nearing Co., Inc.*)

The evan-span, curved-eave Orlyt greenhouse becomes the center of attraction in a garden. The model shown is 14 feet wide, 18 feet long, and over 9 feet high at the ridge. It easily accommodates three tables the length of the greenhouse. (*Courtesy Lord & Burnham*)

order, larger kits include only what you want. For example, if you want to erect a lean-to in the space between two wings of your house and only need the side of the lean-to, you don't have to buy the ends that are normally included.

In style, they range from plain to elegant. Particularly attractive are the models with curved glass eaves. Prices run from about $12 per square foot of floor area for smallest models down to $6 for the largest. Yes, there are economies of scale here.

Glass-to-ground greenhouse models give growing space under the benches and have a lighter, more open appearance than a conventional glass greenhouse built on a masonry wall. But the hot-water radiator heating shown would have to be replaced with a hot-air heating system, if you wanted to use the space under the bench. Roll-up aluminum shades provide protection from the summer sun. The model shown is a curved-eave Orlyt, 10 feet wide, 18 feet long, and almost 10 feet high at the ridge. (*Courtesy Lord & Burnham*)

A custom-designed lean-to provides an unusual amount of headroom for tall plants. (*Courtesy Lord & Burnham*)

This redwood-framed glass green-house is designed for the serious hobbiest. Aluminum glazing bars eliminate the need for putty and make glass replacement easy. (*Courtesy Texas Greenhouse Co.*)

These two photos show installation options for the Elite line of greenhouse from Texas Greenhouse. These Elites feature a galvanized steel subframe that is narrow enough to let in maximum amounts of weak winter sun, yet strong enough to support the whole shell. Since none of the glazing bars are load-bearing, they don't sag or warp—chief causes of sticky vents and doors, as well as leaky weatherseals. (*Courtesy Texas Greenhouse Co.*)

This Series 1200 greenhouse has a cover of rigid polyboard, providing improved insulation. The model shown is 7 × 5 feet. (*Courtesy Sun/America Corp.*)

Janco "Chesapeake" greenhouses are available in even-span and lean-to models in a range of standard sizes, plus modular larger sizes. The model shown is 18 feet wide, 31 feet long, and over 9 feet high at the ridge. (*Courtesy J. A. Nearing Co., Inc.*)

This redwood-framed lean-to comes prefabricated in large sections. The model shown has glass lights and is 11 × 14 feet. This style is also available in other sizes, and there are gable models. (*Courtesy Sturdi-built Manufacturing Co.*)

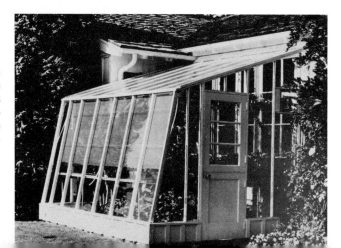

CONSUMER'S GUIDE TO KITS

Consider the following points when choosing a kit greenhouse.

- The range of selection for non-glass greenhouses is limited in relation to the range of glass-covered greenhouses available.
- Greenhouses covered with polyethylene film are made in all sizes from very small patio models to huge commercial models.
- A greenhouse covered with a vinyl (PVC) film costs more initially than one covered with polyethylene film. Yet even with vinyl's longer cover life, it may not cost less in the long run.
- A kit greenhouse covered with flat, rigid plastic costs almost as much as economy glass-covered models.
- Lean-tos tend to cost more per square foot of floor area than do free-standing even-span models. Why? I don't know.
- When similar materials are used, large greenhouses cost less per square foot of floor area than small ones.
- When comparing prices of lines of greenhouses, check what ac-

This window has thermally efficient, two-layer polyboard ends and top, and a clear plastic front. (*Courtesy Grow House Corp.*)

cessories are included in the price, such as benches and ventilators.

● Some greenhouse prices include shipping, freight prepaid. But most makers expect you to pay shipping costs on delivery. When comparing two lines of greenhouses, be sure to consider freight costs, which are based on weight and distance.

● If you know what you want, you can minimize overall freight charges if you buy accessories at the same time you buy the kit—rather than later.

GREENHOUSE KITS—WHO MAKES WHAT

Company	Cover Material				
	Polyethylene Film	Vinyl, or Other Film	Flat FRP or Other Rigid Plastics	Corrugated FRP	Glass
Aluminum Greenhouses, Inc.					X
Baco Leisure Products, Inc.					X
W. Atlee Burpee Company					X
Casaplanta		X			
Clover Garden Products	X				
Dome East Corporation		X			
Grow House Corporation			X		
J. A. Nearing Co., Inc. (Janco)					X
Lord & Burnham					X
McGregor Greenhouses				X	
National Greenhouse Company	X			X	X
Peter Reimuller	X		X	X	
Rough Brothers					X
Solar Technology Corporation					X
Sturdi-Built Manufacturing Company			X		X
Sun America Corporation		X			
Texas Greenhouse Company, Inc.				X	X
Turner Greenhouses	X			X	
Vegetable Factory, Inc.			X		

- When buying a wood-framed kit, be sure the wood is pre-cut and ready for assembly. Predrilled in wood kits is convenient, but not as important.
- If you try to economize by buying redwood framing at a lumber yard rather than with the kit, you may be getting more aggravation than the savings justify. Kits without redwood framing can be good buys, if they include a lot of special metal joints and fasteners and *if you have the tools for cutting lumber to dimension.*

Size and Style							
Small				Large			
Gable	Lean-To	Dome	Other	Gable	Lean-To	Dome	Company
X	X			X	X		Aluminum Greenhouses, Inc.
X	X			X	X		Baco Leisure Products, Inc.
X	X						W. Atlee Burpee Company
X							Casaplanta
X				X			Clover Garden Products
		X				X	Dome East Corporation
X	X		X				Grow House Corporation
X	X		X	X	X		J. A. Nearing Co., Inc. (Janco)
X	X			X	X		Lord & Burnham
X							McGregor Greenhouses
X	X			X	X		National Greenhouse Company
X		X					Peter Reimuller
				X	X		Rough Brothers
	X						Solar Technology Corporation
X			X				Sturdi-Built Manufacturing Company
X							Sun America Corporation
X	X			X	X		Texas Greenhouse Company, Inc.
X	X			X			Turner Greenhouses
X	X						Vegetable Factory, Inc.

● A heater is a heater. You can buy it from major mail order chains or at local stores. Also check local garden stores for greenhouse accessories before ordering by mail.

BUILDING A KIT

The most important two steps in assembling any greenhouse kit are (1) reading the directions, and (2) following them. Foundations for

KIT ASSEMBLY PROCEDURES

A A typical greenhouse kit arrives, as shown, in boxes. Here one long box contains aluminum extrusions. Another box holds panels and hardware. And there are four cartons of glass. The total weight in this instance was 320 pounds.

B This is 2 × 6 redwood construction heart lumber for the base, with half-lapped joints at the corners. Lumber is anchored to concrete piers going 16 inches deep. The piers are wider at the bottom than at the top. Here piers are used at corners and midway along each side. (Note: Before cutting the lumber to length, it may be necessary to lay out the kit base or sill plate on the lumber to determine the exact size needed and to be sure that the anchor bolts emerging from the wood do not interfere with placement of the metal sill plate. If the greenhouse is located on grass, turf should be removed for the wood base and greenhouse floor area. The floor can be covered with crushed rock and a brick walkway.)

greenhouses are discussed in Chapter 4. Many greenhouse kit makers softpedal the need for foundations. But unless your greenhouse is to be only a temporary fixture, it needs a concrete, masonry, or anchored lumber foundation. If your kit comes without power ventilation and you plan to install louvers and a blower, the kit should be assembled as specified. Then check to see what you will have to change to accommodate the equipment. Otherwise, you may mistakenly alter kit parts before you've had a careful look at their position and function.

C The galvanized steel base frame (sill plate) is attached to the redwood with screws. Then assembled end frames are fastened to the steel base with self-tapping screws.

D End frames are supported with twine and wood braces until the eave and ridge rails and intermediate frames are fastened with aluminum bolts and nuts.

E Aluminum wall panels bolted between frames provide rigidity and eliminate the need for diagonal bracing. After the frame is completely assembled, all bolts should be checked for tightness. The track for the sliding door is installed next.

F Glass installation begins with the roof ventilator. The vent is propped open by means of a notched bar. The vent is on the side of the roof facing away from the prevailing wind so it doesn't have to be closed during light rains.

G Glazing goes fast, but gloves are needed. Glazing shouldn't be done on a windy day, because wind can whip the glass around. The glass is cushioned and sealed by self-sticking glazing tape laid on the channels. The tape is not stuck to the glass, only to the channel. The weight of the glass pressed against the spongy tape provides a suction seal. Openings in frames are glazed with two overlapping panes of glass, with the upper pane supported by a clip hooked over the top edge of the lower pane (standard greenhouse practice).

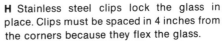

H Stainless steel clips lock the glass in place. Clips must be spaced in 4 inches from the corners because they flex the glass.

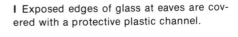

I Exposed edges of glass at eaves are covered with a protective plastic channel.

J The greenhouse shell is now completed and ready for benches, storage, tools, climate-control equipment and plants.

3

Designing and Building Your Greenhouse

The greenhouse frame can be made of wood or aluminum. In my opinion, wood is the better choice because it is the easier to work with. It is usually less expensive, and it is available in a wide variety of sizes, grades, and species. And for small custom greenhouses, wood usually looks better in relation to homes and landscaping.

To me, aluminum has a sterile, hard appearance. It is widely used in kits mainly because it is economical. Kit manufacturers can buy aluminum extrusions "by the mile," to get good prices. When they buy in such quantity, they can also get the exact extrusion cross sections that will save labor in factory assembly.

In planning, you should weigh factors of appearance, likelihood of damage, and costs in your locale.

THE PARTS OF A GREENHOUSE

The names of parts of large commercial greenhouses are fairly well standardized. Kit makers use a variety of terms borrowed from home construction and elsewhere. The following list contains names of parts for commercial greenhouses, with any home-building equivalents noted in parentheses.

Bar: Any greenhouse framing member that receives and supports the edge of the glazing.

Oftentimes, there is no greenhouse kit that will fit a special site. If this is the case, you must either design your own or hire an architect. This custom lean-to frame in front of a ground-level arch makes an unusual greenhouse. (*Adolf deRoy Mark, Architect*)

Corner Bar (corner post): Vertical or near-vertical framing member at a greenhouse corner that supports the weight of the greenhouse roof and also receives vertical edges of the side and gable end glazing.

Drip Channels: Channels on the indoors portions of roof framing bars designed to drain condensed water to the ground rather than allowing it to drip onto vegetation or people. Drip channels are usually molded to the roof bar itself.

Eave: The horizontal framing member of a greenhouse that ties the roof and side bars together and supports the lower edge of the lowest roof glazing light and the upper edge of the highest side glazing light. Overhang may be minimal. In some aluminum frame designs, the eave is omitted.

Gable End Bar (joist): Vertical framing member of the gable that receives and supports the gable glazing.

Girt: A horizontal member of a greenhouse side wall to which the centers of corrugated and flat, fiberglass reinforced plastic (FRP) are fastened to increase rigidity.

Gutter (similar function, though structurally different from house gutter): The horizontal structural member forming the valley between

GREENHOUSE PARTS

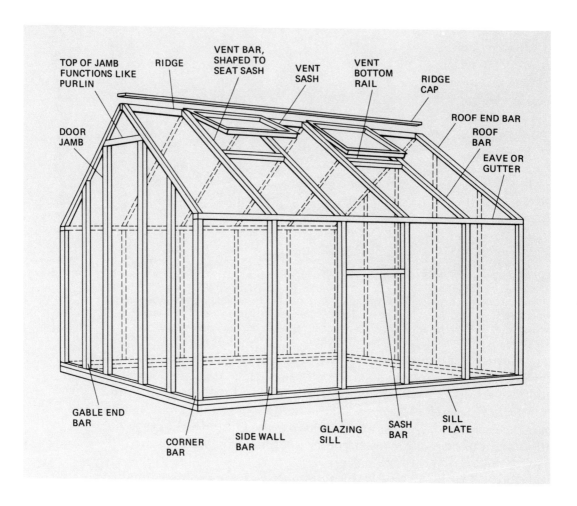

two joined eaves of side-by-side greenhouses. The term is used in both wood and extruded aluminum frame construction.

Purlin (purlin): A horizontal member of a greenhouse roof frame that ties roof bars together. Purlins do not receive the edges of glazing.

Ridge (ridge board): The frame member that joins the top ends of roof bars. A ridge receives the top edge of the glazing. The ridge may also accommodate the vent sashes and hinges.

Roof Bar (rafter): One of a series of parallel structural members of a sloping greenhouse roof supported at the top end by the ridge and at the bottom end either by a vertical or near-vertical member called a side bar, or by a horizontal member called an eave or a gutter. In aluminum construction, the side bar may be a continuation of the roof bar extrusion, notched and bent.

Roof End Bar (gable rafter): The end roof bar of a greenhouse roof. If these are aluminum extrusions, their cross-section is usually quite different from the other roof bars.

Securing Rail: A rail, either wood or aluminum, located near ground level on the outside of a film-covered greenhouse for attachment of the cover edge. If wood, the securing rail may be nothing more than a 2 × 4 or 1 × 3 around which the film is wound before the rail is nailed to the frame.

Side Wall Bar (joist): Vertical or near vertical framing member between the eave and the sill that supports the weight of the roof and receives vertical edges of the glazing.

Sill (sill plate): The framing member of any greenhouse that rests on and is attached to the foundation.

Socket Rail: Vent sashes in aluminum-framed greenhouses are often attached with socket hinges rather than with butt hinges, which are harder to make watertight. The hinge socket is formed as part of the extrusion. (Note: Some aluminum greenhouse extrusions have similar-appearing C-shaped sockets formed in them to receive self-tapping screws.)

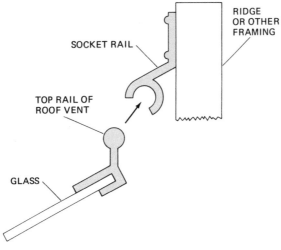

SOCKET RAIL ASSEMBLY

Vent Bar: A bar shaped on one or both sides to seat a vent sash.

Vent Bottom Rail: A horizontal framing member used either on the roof, side wall or end wall of a greenhouse to support and provide a weather-tight seal for a vent.

GREENHOUSE COVERS

Designing and building a home greenhouse starts with the selection of cover material. Your choice of glass, acrylic, film, or fiberglass reinforced plastic (FRP) will determine structural and glazing requirements. A glass cover needs frames spaced 16 to 24 inches apart with the sides of the frames shaped to receive the edges of the glass lights. A flat acrylic or FRP cover can be used with frames spaced and designed for glass. Or frames can be farther apart if cross supports are added to keep the larger and thinner plastic panels from flexing. Flat acrylic and FRP sheets can also be mounted on top of the frame members, eliminating a lot of work otherwise required to form the frames so that lights will fit well. Corrugated FRP panels in very large pieces can be mounted on top of or outside the frame. The frame structure for a film-covered greenhouse can be comparatively light and made of wood, aluminum, extruded plastic, or thin-wall electrical conduit. A polyethylene film cover can be mounted over almost any kind of a frame, provided there are no sharp edges and there is a means of tying down the edges.

Glass

Five kinds of glass are used in greenhouses: SSB, DSB, wire-reinforced, laminated, and tempered. SSB is ordinary single-strength window glass. It is $\frac{3}{32}$-inch thick and is not as strong as double-strength window glass (DSB) which is $\frac{1}{8}$-inch thick. Note that DSB is double-strength, not double-thickness. Panes of glass are called *lights*, or *lites*. Although SSB and DSB glass is manufactured in sizes up to 48 × 84 inches, the largest size that should be considered for a greenhouse is 24 × 24 inches.

Whether or not single-strength glass is adequate for a backyard greenhouse is the subject of debate. Commercial greenhouse operators use double-strength glass.

Wire glass has a wire mesh or screen sandwiched into it during manufacture. When the glass is struck hard enough to break it, the wire allows only small chunks rather than large dangerous pieces, to break free. If your glass greenhouse is to be glazed all the way to ground level, some building codes require that the lower glass be either wire-reinforced or tempered for extra strength. Wire glass is the more expensive.

Laminated glass is a combination of two (or more) lights of glass with a tough, transparent vinyl layer heat-sandwiched between the glasses to form a single construction. Many combinations are made to meet various glazing requirements and building codes. Of primary interest in greenhouse construction is a $\frac{15}{64}$-inch safety glass, such as PPG Industry's Duolite. This glass consists of two pieces of $\frac{7}{64}$-inch sheet glass laminated with a vinyl interlayer.

If you have a glass light on and around the door of your greenhouse, it must be safety glass, usually tempered glass. Tempered glass is four to five times stronger than ordinary annealed glass of the same thickness, but it is breakable. When it does break, however, it shatters into small, relatively harmless, cubical fragments rather than dangerous blades and shards. The tempering process is done after the glass has been cut to size. Tempered glass is made by subjecting annealed glass to a heat treatment that gives the glass increased mechanical strength and resistance to thermal stress. The minimum thickness of tempered glass is $\frac{3}{16}$-inch. A light of tempered glass can't be cut to fit; it can be used only in standard manufactured sizes.

All glass can be broken when struck by hard objects or strong winds. So it must be handled carefully during the installation. You should never attempt greenhouse glazing on a windy day. For all its drawbacks, glass is cheap, and widely available in exactly the size you want. It is noncombustible, and for all practical purposes it has unlimited life, barring accident.

Single-strength (SSB) and double-strength (DSB) glass are packed in boxes that will hold 50 square feet of glass. The number of lights per box depends on the size of each pane, with the number adjusted to yield close to 50 square feet of glass. Boxes of SSB glass weigh about 70 pounds, DSB 90 pounds. You should shop around locally for a reasonable box price. Also before you start designing your greenhouse, determine what sizes your local dealer stocks regularly so that if you need replacement panes, you will be able to get them without delay.

STANDARD GLASS SIZES AND PRICES

Size	Lights per Box	Price per Box	
		Single Strength	Double Strength
10 × 14″	51	$15.00	$22.00
16 × 18″	25	15.00	22.00
16 × 24″	19	15.00	22.00
18 × 20″	20	15.00	22.00
20 × 20″	18	15.00	22.00
20 × 24″	15	15.50	27.00
20 × 30″	12	15.50	27.00
24 × 26″	12	16.10	36.00
24 × 30″	10	16.10	36.00

Bar-cap Glazing

Aluminum bar-cap glazing provides the best protection against leaks, and protects the glazing compound or plastic sealing strip from exposure to sunlight that will cause deterioration. Bar caps can be used with either wood or aluminum frame construction. The design of bar caps prevents any glass slippage because each bar cap is the same length as the light, and holds and supports the light for its entire length, not in just one or two places, as with clips. And the upper end of the bar cap acts as a stop for the light above.

Installation is simple; sealant is gunned onto the *glazing* bar shoulder, or taped sealant is laid on. The light is pressed into place, a bead of sealant is gunned to seal the top edge of the light and the glazing bar, the bar cap is laid over the light and fastened with screws onto the aluminum or wood frame member.

When designing your own greenhouse, dimension your framing so it can be fitted with commercially available bar caps. These bar caps come in three sizes—18-inch, 20-inch, and 24-inch. These cost $.20 to $.30 each, depending on length. Those supplied by National Greenhouse Company fit frames with $\frac{5}{8}$- or $\frac{11}{16}$-inch tongues.

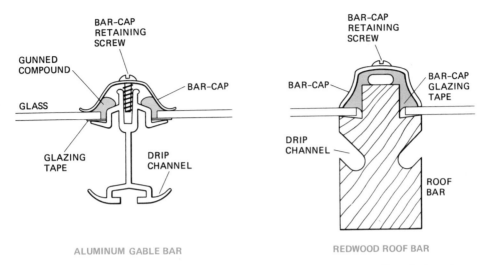

ALUMINUM GABLE BAR

REDWOOD ROOF BAR

These drawings show cross sections of two popular types of bars as well as glazing elements. At left is an aluminum bar used by Lord & Burnham. At right is a redwood bar used by National Greenhouse.

Vinyl-spline Glazing

Vinyl spline glazing is commonly used in aluminum storm doors. It can only be used with aluminum framing extrusions specifically designed for it. Sealant is gunned onto the glazing bar shoulders; the light is pressed in place; and flexible vinyl spline is pressed over the glass and locked into a groove in the glazing bar. No additional gunned sealant is used on top of the light. Although vinyl does not deteriorate when exposed to the elements and sunlight, it doesn't provide as much protection for the glazing compound used to bed the light.

Stainless steel clips provide an inexpensive way to secure lights to glazing bars. There are several clip designs in use and each must be used with the glazing bar extrusion for which it is designed. Foam tape sealing strips are applied to the glazing bar shoulders. Even though this tape has an adhesive on only one side for attachment to the glazing bar, a good seal can be made. Two clips on each side are snapped into place 4 to 6 inches from the top and bottom edges of the light. Support for lapped lights is obtained by using S-shaped clips that hook over the top edge of the lower light.

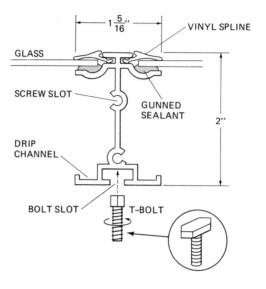

This cross section shows the spline glazing used by the J A Nearing Co.

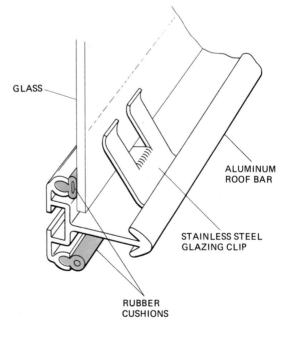

Here is the clip glazing employed by the British Aluminum Company.

This vinyl H-shaped came is similar to the lead came used in stained-glass work. It allows glass panes to be butted instead of lapped on the sides of a greenhouse. And it prevents entry of dust and water. (*Courtesy Texas Greenhouse Co.*)

Some people claim glazing clips do not provide enough support to keep glass from slipping under heavy snow loads. The glass in my own greenhouse is clip-supported, and I have had no slippage in three winters. Clip-supported glass has another disadvantage though: When used with lapped lights, and to a lesser extent with single lights, the clips need to flex the glass to hold it in place. Thus, the flexed glass lights leave gaps that allow water leaks. S-clips for lapped glass keep the lights apart by the thickness of the clip metal. Gaps are a major cause of winter heat loss.

Acrylic

Acrylic is a rigid, resilient plastic that is available in colorless sheets or sheets with transparent tints, translucent and opaque colors, and textured surfaces. Trade names include Plexiglas (Rohm & Haas) and Lucite (American Cyanamid).

Unlike plywood which comes in exactly cut 48 × 96-inch sheets, acrylic sheets are oversize—a minimum of 2.5 percent. All of the thicknesses listed in the accompanying table on acrylic sheets are available in 36 × 48-inch, 36 × 60-inch, and 48 × 72-inch nominal standard sheets.

"Nominal" here means that sizes are approximate. The thicknesses are also nominal, and they can vary from minimum to the maximum tolerances from one end of a sheet to the other. This can present a problem if you are planning to insert the acrylic into slots.

ACRYLIC SHEET STANDARD SIZES
(Rohm & Haas Plexiglas)

Size	Thickness in mils						.125 .187 .250
	.030	.040	.050	.060	.080	.100	
36 × 48"	X	X	X	X	X	X	X
36 × 60"				X	X	X	X
36 × 72"				X	X	X	X
42 × 84"							X
48 × 48"							X
48 × 60"							X
48 × 72"							X
48 × 96"							X

Acrylic sheet is made in several grades. But only two are of interest for greenhouses. One type is safety glazing (Rohm & Haas Plexiglas K). The other is decorative safety glazing (Rohm & Haas Plexiglas G). Grade G can be heat formed and costs 50 percent more than Grade K.

You can buy safety glazing acrylic sheets from retailers or glass dealers already cut into standard sizes or cut to your specifications. You can save money by buying the plastic in standard sheets and cutting it yourself.

Sawing acrylic sheet is fairly easy. Leave the protective paper on. If there is no paper on the sheet, apply 2-inch masking tape along the cut line on both sides of the sheet. The tape keeps hot chips from getting into the kerf behind the blade and rewelding. If you use a saber saw, use a 32-tooth metal cutting blade and a slow feed. If you want to use a table saw or radial-arm saw, use a blade designed for acrylic or else a fine-tooth veneer blade. You can also cut acrylic sheet by scribing and breaking just as you would cut cut glass, but you must use a scriber made for acrylic, and scribe the line several times, rather then just once as for glass.

Glazing a sash with acrylic requires extra allowances in the sash for expansion and contraction of the acrylic as it changes temperature.

Acrylic sheet (Plexiglas) can be cut with saber, scroll, or table saws. Sheets up to $\frac{1}{4}$-inch thick can be scribed with a special carbide tool and then broken. Using a straight edge as a guide, place the point of the scribing tool at the edge of the material. Then, while applying firm pressure, draw the cutting point the full length of the cut. You should repeat this 5 to 6 times for $\frac{1}{16}$- to $\frac{3}{16}$-inch material and 7 to 10 times for $\frac{1}{4}$-inch. Leave the masking paper or tape on the Plexiglas during the cutting and breaking. (*Photos courtesy Rohm & Haas*)

To break the material, position the scribed line face up over a $\frac{3}{4}$-inch diameter wood dowel that runs the length of the intended break. Hold the sheet with one hand and apply downward pressure on the short side of the break with the other. Keep your hands adjacent to one another and successively reposition them about 2 inches behind the break as it progresses along the scribed line. The minimum feasible cutoff width is about $1\frac{1}{2}$ inches. Patterned Plexiglas cannot be scored and broken.

This is not necessary with glass. Thus large sashes for acrylic must have a deeper bite than sashes for glass to accommodate the expansion and contraction.

THERMAL EXPANSION FACTORS

Side Dimension	Clearance Required
12–36 inches	$\frac{1}{16}$ inch
36–48 inches	$\frac{1}{8}$ inch
48–60 inches	$\frac{3}{16}$ inch

Rigid, flat-sheet acrylic or fiberglass can be substituted for glass when bar caps are used. Plastic sheet can also be substituted when vinyl splines are used, provided the plastic is at least as thick as single-strength glass. Yet this is not likely to be done because both types of plastics are generally used in thinner sheets, owing to their greater strength and higher cost. Neither plastic is sufficiently rigid to be secured with spring glazing clips.

Rigid plastic covers are usually more successful when applied to the outside of the frame. Some aluminum framing and steel pipe framing used in large commercial greenhouses is designed for corrugated fiberglass covers. Wood frames are used almost exclusively for home greenhouses with rigid plastic covers.

Remember to take care in fastening the plastic to the framing because nail and screw holes tend to leak. For this reason, special nails and wood screws equipped with neoprene washers are used to ensure that the hole in the plastic is sealed. Always pre-drill nail or screw holes; the holes should be oversize to allow for expansion and contraction of the plastic. For best overall economy and to minimize the number of potential leaks in joints, plan to use sheets of plastic in standard sizes, and plan for a minimum of cutting.

Fiberglass Reinforced Plastic (FRP)

These panels are made both flat and corrugated. Corrugated panels are usually preferred over flat in new construction because cor-

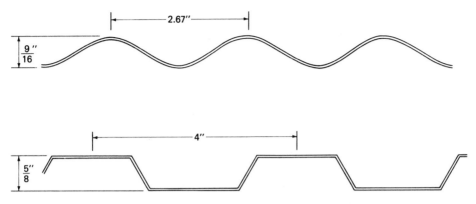

These are cross sections showing dimensions of corrugated fiberglass-reinforced plastic.

rugated panels can span longer distances (not wider distances) without support. Accessories are available to fit and seal the corrugated sheets against flat framing members. There are several styles of corrugations.

FRP panels are clear or tinted. Owens-Corning recommends that only clear untinted FRP panels be used for plant propagation. Clarity varies with the composition of the polyester resin and depends on how close the light-refraction indices of the resin and glass fibers match.

The maximum allowable spacing of purlins and girts depends on the dimensions of the FRP panel corrugations, as shown below.

Corrugation	Purlin Spacing (Roofs)	Girt Spacing (Side Walls)
2.25-inches	32 inches	44 inches
2.5 inches	48 inches	60 inches
2.67 inches	54 inches	66 inches

When installing FRP panels, begin at the lee end of the greenhouse and work towards the windward end. This will reduce air and rain infiltration through the lapped joints from the prevailing wind.

Overlap one corrugation on roofs when using universal vinyl lap seal. Overlap 1⅓ corrugations when using clear mastic seal. On side-walls, overlap one corrugation with any sealant. Endlap on roofs with a pitch of less than 4 inches in 12 should be 8 inches. For steep roofs, a 6-inch endlap is adequate. Fasten all panels to the greenhouse structure through the valleys of every second corrugation with either nails or screws equipped with neoprene washers. First, drill pilot holes.

Polyethylene Film

Ultraviolet-inhibited polyethylene film, such as Monsanto's 602, is the least expensive greenhouse cover, even taking into consideration annual replacement, which is not always necessary. A double layer can be applied, and the space between can be pressurized with a small blower to provide effective "double glazing" and significant savings in fuel cost.

A big advantage of polyethylene film over other films besides low cost is that it can be purchased in very large pieces, enabling you to cover a greenhouse with very few pieces. This not only makes recovering easier, but also eliminates seams through which warm indoor air can escape.

POLYETHYLENE FILM, SIZES AVAILABLE (MONSANTO 602)

Mil	Size	Put Up	Square Feet per Roll	Approximate Gross Weight per Roll
4	10 × 100'	Flat sheet	1000	21
4	12 × 100'	Flat sheet	1200	26
4	14 × 100'	Flat sheet	1400	28
4	16 × 100'	Flat sheet	1600	32
4	20 × 100'	Centerfold	2000	43
4	24 × 100'	Centerfold	2400	52
6	14 × 100'	Centerfold	1400	45
6	16 × 100'	Centerfold	1600	51
6	20 × 100'	Centerfold	2000	64
6	24 × 100'	Centerfold	2400	76
6	32 × 100'	Gusseted	3200	105
6	40 × 100'	Gusseted	4000	133

Double Glazing

Fuel costs are rising so high that home greenhouse operators in many parts of the country are choosing between double-glazing their greenhouse and shutting down in winter. Any greenhouse frame design must be approached with the consideration of how it can be double glazed, either at the outset or later.

Fortunately, an existing greenhouse with single-pane glass doesn't have to be double glazed with glass. Plexiglas can be used. Or you can tie a large piece of polyethylene film over the outside; this may not be attractive, but it is effective. The Plexiglas should be used in .06- or .08-inch thicknesses depending on availability. These are the lowest-priced thicknesses; thinner and thicker types both cost more. In wood-framed greenhouses, the plastic can be attached to the inside surface of the frame with wood screws and flat washers, or with wood screws and wood battens. For aluminum frames with T-bolt slots, T-bolts and washers can be used. For aluminum extrusion frames, such as used in the Arrow greenhouse, the frames must be drilled for self-tapping screws.

Two kit greenhouses that use rigid plastic covers in unusual ways are worth noting. Vegetable Factory greenhouses are framed with aluminum extrusions, but the cover is made up in panels consisting of two layers of translucent fiberglass cemented to peripheral aluminum I-beams. This provides considerable rigidity and thermal insulation. Grow House brand greenhouses are covered with a double-layered

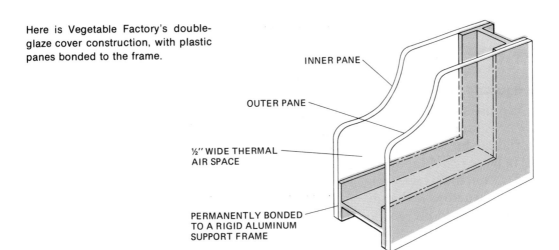

Here is Vegetable Factory's double-glaze cover construction, with plastic panes bonded to the frame.

INNER PANE

OUTER PANE

½" WIDE THERMAL AIR SPACE

PERMANENTLY BONDED TO A RIGID ALUMINUM SUPPORT FRAME

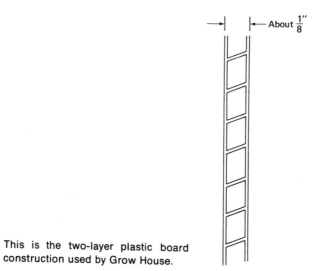

About $\frac{1}{8}$"

This is the two-layer plastic board construction used by Grow House.

translucent plastic that is rigid and internally ribbed. It resembles corrugated cardboard. The ribs are claimed to act as louvers that distribute sunlight evenly inside the greenhouse, and the ⅛-inch dead-air space is claimed to provide excellent insulation. This later claim, of course, would only be valid if the open edges of the plastic sandwich were sealed to trap the air. A greater separation of the two skins would also provide better thermal insulation.

GREENHOUSE FRAMING

Aluminum

There are two ways to design an aluminum-framed greenhouse. You can use do-it-yourself shapes and common structural shapes, or you can use special greenhouse shapes. In the first approach, the cover material can be anything except glass—that is, anything that can be used in large panels and fastened to the outside surface of the framing. An exception to this would be a composite construction—an aluminum extrusion structure with redwood roof, and framing milled to receive glass. This would be practical but expensive.

As far as I know, the only company that sells special greenhouse

extrusions separately is National Greenhouse. The extrusions are from
their Eaglet line. One problem here is that you would have to use the
Eaglet eave and ridge angles or else the parts wouldn't go together.
But such a project would be interesting. (*Text continues on page 82.*)

These are aluminum extrusions used in National Greenhouse's "Eaglet" line. Weights
given are pounds per foot. Extrusions are priced by the pound.

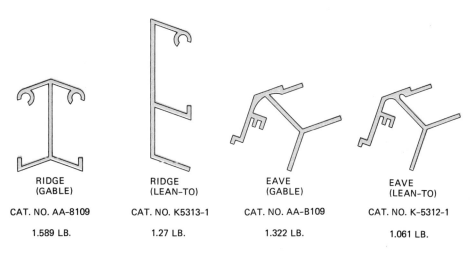

RIDGE (GABLE)	RIDGE (LEAN-TO)	EAVE (GABLE)	EAVE (LEAN-TO)
CAT. NO. AA-8109	CAT. NO. K5313-1	CAT. NO. AA-B109	CAT. NO. K-5312-1
1.589 LB.	1.27 LB.	1.322 LB.	1.061 LB.

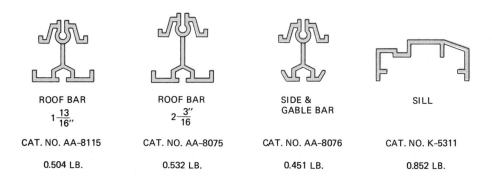

ROOF BAR $1\frac{13}{16}''$	ROOF BAR $2\frac{3}{16}''$	SIDE & GABLE BAR	SILL
CAT. NO. AA-8115	CAT. NO. AA-8075	CAT. NO. AA-8076	CAT. NO. K-5311
0.504 LB.	0.532 LB.	0.451 LB.	0.852 LB.

STANDARD INDUSTRIAL ALUMINUM SHAPES

STRUCTURAL ANGLE
EQUAL LEGS

A	B		A	B
$\frac{3}{4}$ × $\frac{3}{4}$			3 × 3	
1 × 1			$3\frac{1}{2}$ × $3\frac{1}{2}$	
$1\frac{1}{4}$ × $1\frac{1}{4}$			4 × 4	
$1\frac{1}{2}$ × $1\frac{1}{2}$			5 × 5	
2 × 2			6 × 6	
$2\frac{1}{2}$ × $2\frac{1}{2}$			8 × 8	

STRUCTURAL ANGLE
UNEQUAL LEGS

A	B		A	B
1 × $\frac{3}{4}$			$2\frac{1}{2}$ × $1\frac{1}{4}$	
$1\frac{1}{4}$ × $\frac{3}{4}$			$2\frac{1}{2}$ × 2	
$1\frac{1}{2}$ × $\frac{3}{4}$			3 × 2	
$1\frac{1}{2}$ × 1			$3\frac{1}{2}$ × $2\frac{1}{2}$	
$1\frac{1}{2}$ × $1\frac{1}{4}$			4 × 3	
2 × $1\frac{1}{2}$			& LARGER	

EXTRUDED ANGLE
EQUAL LEGS

A	B		A	B
$\frac{1}{2}$ × $\frac{1}{2}$			2 × 2	
$\frac{5}{8}$ × $\frac{5}{8}$				
$\frac{3}{4}$ × $\frac{3}{4}$				
1 × 1				
$1\frac{1}{4}$ × $1\frac{1}{4}$				
$1\frac{1}{2}$ × $1\frac{1}{2}$				

EXTRUDED ANGLE
UNEQUAL LEGS

A	B		A	B
$\frac{3}{8}$ × $\frac{3}{4}$			1 × 2	
$\frac{1}{2}$ × 1			1 × 3	
$\frac{1}{2}$ × $1\frac{1}{4}$			$1\frac{1}{2}$ × $3\frac{1}{2}$	
$\frac{3}{4}$ × 1			$2\frac{1}{2}$ × $5\frac{1}{4}$	
$\frac{3}{4}$ × $1\frac{1}{2}$				
1 × $1\frac{1}{2}$				

STRUCTURAL CHANNEL

A	B
3 × 1.41
3 × 1.5
3 × 1.6
4 × 1.58
4 × 1.65
4 × 1.73
& LARGER

EXTRUDED CHANNEL

A	B		A	B
$\frac{1}{2}$ × $\frac{3}{8}$			$1\frac{1}{2}$ × $\frac{3}{4}$	
$\frac{1}{2}$ × $\frac{1}{2}$			$1\frac{1}{2}$ × $1\frac{1}{2}$	
$\frac{1}{2}$ × $\frac{3}{4}$			2 × $\frac{1}{2}$	
$\frac{3}{4}$ × $\frac{3}{4}$			2 × 1	
1 × $\frac{1}{2}$			2 × 2	
1 × 1			MANY OTHER	
$1\frac{1}{2}$ × $\frac{1}{2}$			SIZES	

Many of the sizes shown are available with different web thicknesses. Structural shapes come in 25-foot lengths; extruded shapes come in 16-foot lengths.

STRUCTURAL H-BEAM	STRUCTURAL I-BEAM	STRUCTURAL WIDE-FLANGE BEAM
A B	A B	A B
4 X 4	3 X 2.3	6 X 4
5 X 5	3 X 2.5	6 X 6
6 X $3\frac{1}{3}$	4 X 2.7	& LARGER
6 X 6	4 X 2.8	
8 X 8	& LARGER	

EXTRUDED TEE		STRUCTURAL TEE		STRUCTURAL ZEE
A B	A B	A B	A B	A B
$\frac{3}{4}$ X $\frac{3}{4}$	1 X $\frac{3}{4}$	2 X 2	4 X 3	1 X 1
$\frac{3}{4}$ X 1	1 X 1	$2\frac{1}{4}$ X $2\frac{1}{4}$	4 X 4	$1\frac{1}{2}$ X $1\frac{1}{4}$
$\frac{3}{4}$ X $1\frac{1}{4}$	$1\frac{1}{4}$ X $\frac{7}{8}$	$2\frac{1}{2}$ X $2\frac{1}{2}$		$1\frac{1}{2}$ X $1\frac{1}{2}$
$\frac{7}{8}$ X $1\frac{1}{4}$	2 X $\frac{3}{4}$	3 X 3		
1 X $\frac{1}{2}$	2 X 2			

Wood

For greenhouse framing, wood has advantages: It is readily available in many usable shapes. It is relatively inexpensive compared to metal. And it is easily worked with tools and techniques common to home workshops. But wood has disadvantages too. To be as strong as metal, wood members are larger and block more light. Wood needs periodic applications of protective finishes, unless redwood heartwood is used. And wood is susceptibile to rot and termites, again unless redwood heart is used. Wood also has a more traditional look, as compared to bright shiny aluminum. Because of wood's insulating value, wood-framed greenhouses may lose 5 to 8 percent less heat than metal-framed greenhouses of like size.

For a permanent greenhouse, redwood should be used for all

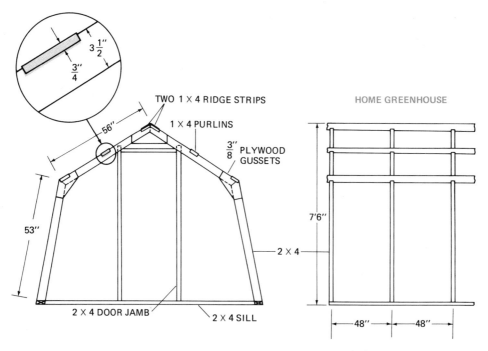

The Home Greenhouse can be sized to suit your needs. Using only half of it, you have a lean-to. Ridge strips and purlins should be inletted into the 2 × 4 rafters, and they'll support either film or a corrugated-plastic cover. The plan is available as Circular 880 from the U. of Illinois College of Agriculture, Cooperative Extension Service, Urbana, IL 61801, and as Plan 6181 from the U. of Maryland, College Park, MD 20742.

structural parts. Not all parts of the redwood log are resistant to decay and termites; only the dark-colored heartwood is. This heartwood is sold as Clear All Heart and Construction Heart. Clear All Heart is normally sold kiln-dried and contains no knots and no visible defects, but it is priced accordingly. Construction Heart contains tight knots which might be sound or unsound, and encased knots, and is sold unseasoned.

Permanence in greenhouse building is relative. A greenhouse built with ordinary construction lumber and exterior-grade plywood can last many years even without careful upkeep. The frame of a film-covered greenhouse is easily accessible for painting and repair whenever the cover is replaced.

Using construction lumber (dimension lumber) in small greenhouse frames usually results in framing that is bulkier than is structur-

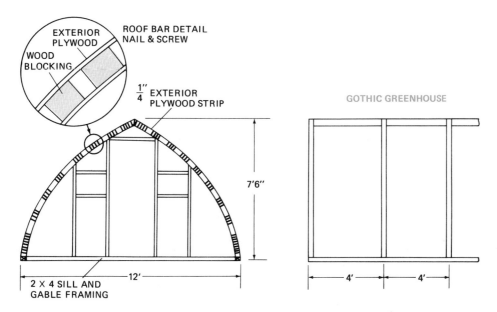

The curved rafters of this Gothic Greenhouse are made by using a jig to sandwich 1-inch blocks between ¼-inch plywood strips. Bottom ends can be bolted to a wood sill or a concrete foundation. For more headroom, the contour of the rafters can be made steeper if you reduce the width of the greenhouse to 10 feet. But the roof curve, itself, is limited to 8 feet (the length of plywood paneling). The plan is available as Circular 487 from Virginia Polytechnic Inst., Blacksburg, VA 24061.

ally necessary. Wood, when used in commercial greenhouse construction, is made up in specially milled shapes. Use of these shapes in smaller home greenhouse designs is quite feasible, although the extra cost may not seem justified to you.

Because of the knots in dimension lumber, it's not practical to cut down 2 × 4s and 2 × 6s. Clear grades of redwood and clear or select pine should be used. Of course oak or birch might be less expensive and will serve. But redwood is the best wood for greenhouse construction.

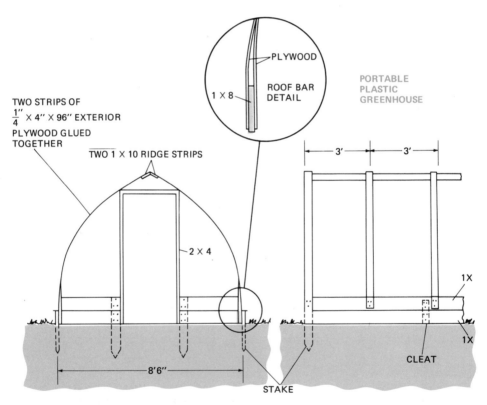

Curves of the rafters of the Portable Plastic Greenhouse are achieved by gluing ¼-inch plywood strips in a jig and sandwiching their bottom ends over the side 1 × 10 wall board, by means of glue and screws. Hardboard could serve as a substitute for the plywood. The roof curve, itself, is limited to 8 feet (the length of paneling). This frame will support only a film cover. Plan 5946 is available from the U. of Maryland, College Park, MD 20742.

MILLED REDWOOD SHAPES (*Continued on next page*)

Milled redwood shapes, shown here and on the next page, are used for a medium-sized greenhouse. Dimensions are approximate. (*Courtesy National Greenhouse Co.*)

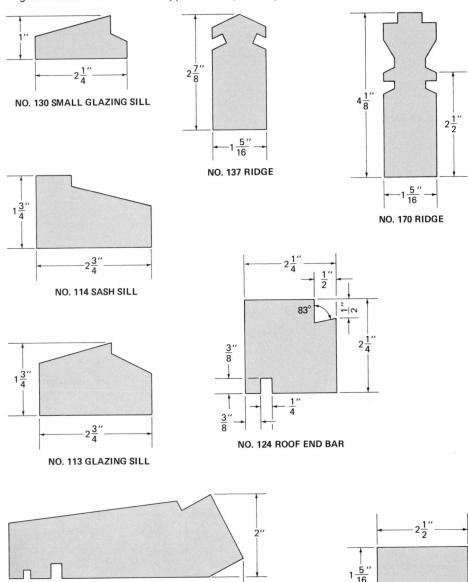

NO. 130 SMALL GLAZING SILL

NO. 137 RIDGE

NO. 170 RIDGE

NO. 114 SASH SILL

NO. 124 ROOF END BAR

NO. 113 GLAZING SILL

NO. 100 EAVE PLATE

NO. 171A CAP (LEAN-TO)

MILLED REDWOOD SHAPES (*Continued*)

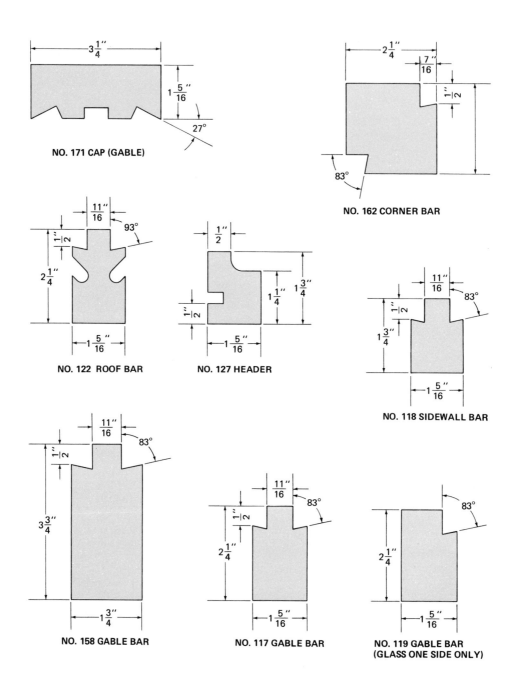

NO. 171 CAP (GABLE)

NO. 162 CORNER BAR

NO. 122 ROOF BAR

NO. 127 HEADER

NO. 118 SIDEWALL BAR

NO. 158 GABLE BAR

NO. 117 GABLE BAR

NO. 119 GABLE BAR
(GLASS ONE SIDE ONLY)

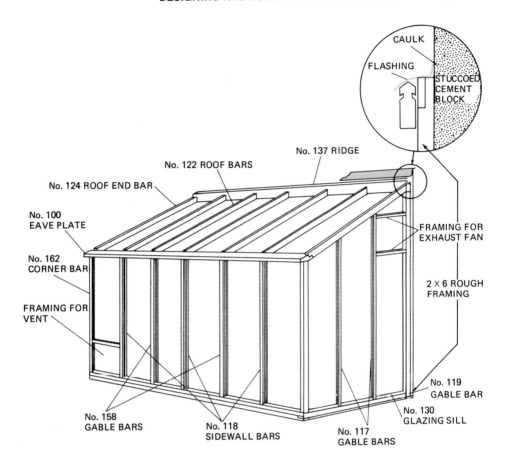

CAULK

FLASHING

STUCCOED CEMENT BLOCK

No. 137 RIDGE

No. 122 ROOF BARS

No. 124 ROOF END BAR

No. 100 EAVE PLATE

No. 162 CORNER BAR

FRAMING FOR VENT

FRAMING FOR EXHAUST FAN

2 X 6 ROUGH FRAMING

No. 119 GABLE BAR

No. 130 GLAZING SILL

No. 158 GABLE BARS

No. 118 SIDEWALL BARS

No. 117 GABLE BARS

Fasteners

Common nails and shiny zinc-plated screws are not satisfactory for exterior or exposed construction of the kind encountered in building greenhouses, gazebos, or other yard structures. For more durable, corrosion resistant hardware you must use fasteners made of stainless steel, aluminum alloy, and hot-dipped galvanized metal. Plain steel nails and screws, even zinc-plated screws, will corrode, stain the wood, and fail. Copper and brass fasteners, while corrosion resistant, tend to discolor redwood. Rosined nails are not corrosion resistant.

Wood screws are used to fasten pieces one-inch thick or less and for mounting hinges and other hardware. Lag bolts are used to join larger pieces, and to attach heavy metal angles and channels to wood members.

To avoid splitting the wood, and to obtain joints with maximum strength, predrill holes for wood screws and lag bolts. For both of these fasteners in redwood construction, the hole for the threaded portion of the screw should be about $\frac{7}{8}$ the root diameter of the thread, and $\frac{7}{8}$ the diameter of the shank for that portion. Countersink holes for flat headed screws, and use washers under round heads and lag bolt heads.

Wood screws and lag bolts have different resistances to withdrawal and to lateral loads. Withdrawal resistance varies directly with

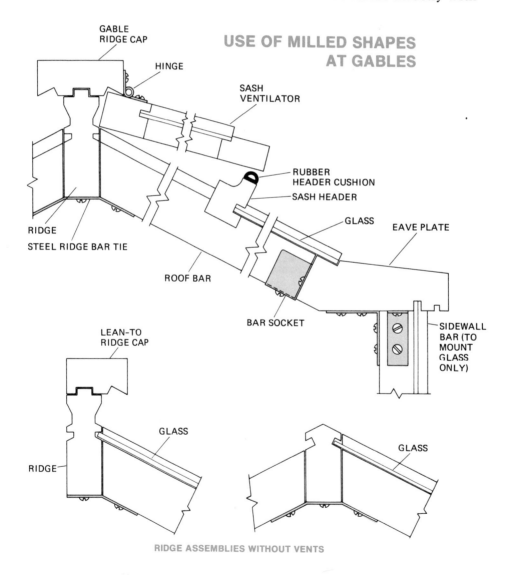

USE OF MILLED SHAPES AT GABLES

RIDGE ASSEMBLIES WITHOUT VENTS

the diameter and the length of the screw. Screws or bolts that are embedded in the wood a distance at least seven times their diameter seem to handle lateral loads well. Generally speaking, withdrawal resistance also varies with hardness of the wood, and withdrawal resistance is

WOOD FASTENERS

Bolts and pins for wood joints require predrilled holes. For maximum strength in hardwoods, the hole should be made about the same diameter as the bolt or pin. But for softwoods, the hole should be slightly smaller so that the bolt or pin must be driven through.

NAILED METAL CONNECTORS

Though more costly than nails alone, metal connectors allow you to frame an entire structure, if you wish, without having to toe-nail into a joint. These devices make strong joints, and they help avoid accidentally marring and splitting the wood.

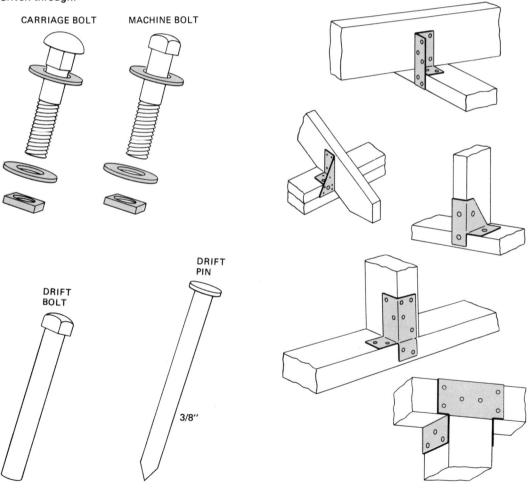

CARRIAGE BOLT   MACHINE BOLT

DRIFT
PIN

DRIFT
BOLT

3/8"

less in end grain than across the grain. For example, in redwood, end-grain withdrawal resistance is only 70 percent of the withdrawal resistance across the grain. The safe lateral loading of wood screws across grain is almost 2½ times that of common nails of the same diameter.

For fastening of wood members together, bolts are primarily used. Bolts are normally best for side grain to side grain, and side grain to end grain, and to fasten metal members to wood side grain. Holes must be predrilled. For maximum joint strength, in softwoods, the hole through both pieces should be slightly smaller than the shank diameter so that the bolt has to be driven through. In hardwoods, the hole should be the same size as the bolt. Washers must be used under the nuts of carriage and machine bolts, and under the heads of machine bolts, to prevent the wood from being crushed when the nut is tightened. Since wood may continue to shrink after assembly, tightness should be checked periodically.

Machine bolts have either square or hex heads. Carriage bolts have slightly domed heads with a square-sectioned shank under the head to prevent the bolt from turning as the nut is tightened. Thus, the bolt's square shank must be fitted tightly into the predrilled hole.

Three other types of fasteners are used in post and beam construction. These are drift pins, drift bolts, and dowels. They are used primarily to join side grain to end grain and when lateral strength is important. Since they don't have threads, they have poor withdrawal resistance, behaving like oversize, straight-driven nails. A drift bolt is essentially a machine bolt without any threads and a square end; a drift pin has a round head and a pointed end. Both are driven into predrilled holes. A dowel may be wood (then sometimes called a peg) or metal. While a drift must be driven through one member into the other, the dowel may also be "blind," hidden inside the joint. Glued wood dowels produce strong joints that have excellent withdrawal resistance as well as lateral resistance. Dowel joints (metal dowels) are often used to attach posts to concrete.

Nailing is the least satisfactory way of fastening greenhouse and other outdoor structures. Where nails are used, they should be either corrosion-resistant stainless steel, aluminum alloy, monel, hotdipped galvanized, or copper. Aluminum nails are widely available.

Nails used with metal connectors will produce strong joints without the danger of splitting the wood and accumulating unsightly hammerhead imprints. These connectors are made in a variety of forms including post anchors, joist hangers, post caps, and utility plates.

Adhesives

Adhesives for greenhouse construction, and all outdoor building for that matter, must be waterproof. Not just water resistant, but waterproof. Four commonly available adhesives meet this requirement. They are described in the text below and in the accompanying table.

Resorcinol glue resists both heat and water and has high strength. It is widely used in outdoor and marine work. The glue comes in two parts, a dark red liquid resin and a tan powdered hardener. The components are mixed four parts resin to one part hardener by weight. The mixture can be used for gluing for three hours after mixing. The glue will not cure at temperatures below 70° F. Joints must be clamped, with curing times of 10 hours at 70° F, dropping to $2\frac{1}{2}$ hours at 90° F.

Epoxy glues are made in various formulations. All come as two liquids that are mixed just before use. They are heat and water resis-

WATERPROOF ADHESIVES

Name	Put Up	Mixing	Glue Line	Clamping	Disadvantages	Cost per oz. of Mixed Glue
Resorcinol	1-qt. can liquid resin 1-10 oz. can powdered hardener (also put up in other sizes)	Equal parts by weight	Dark red	Pressure required	Messy, stains wood	0.35
Epoxy	$\frac{1}{2}$ oz. tube resin $\frac{1}{2}$ oz. tube hardener	Equal parts by volume	Colorless	Only enough pressure to hold pieces in position	High cost	2.00
	pint can liquid resin pint can liquid hardener (also put up in other sizes)	Equal parts by volume	Colorless	Only enough pressure to hold pieces in position	None	0.47
Filled Epoxy	5 oz. tube resin 5 oz. tube hardener	Equal parts	Gray	Only enough pressure to hold pieces in position	None	0.30

tant, and expensive. Fast-curing formulations have curing times as short as 5 minutes, but the glue must be mixed and the joint closed in one minute. Other types have a working life of up to one hour, and they require a 12-hour cure. Curing is speeded as temperature increases.

Filled epoxy glue is not as expensive as other epoxy glues and can be used for poorly-fitting or rough-surfaced joints. The bond obtained is both water and heat resistant. The glue comes in two tubes; the black and white contents are mixed to obtain a gray-colored glue that cures in 4 to 6 hours at 70° F. You can get this glue from Sears, Roebuck & Company.

STUCCOED-BLOCK LEAN-TO

A greenhouse for a small patio can be simply standard glass sliding doors, stuccoed cement-block side walls, and a roof framed in rough lumber. The cover here is ⅜-inch Plexiglas.

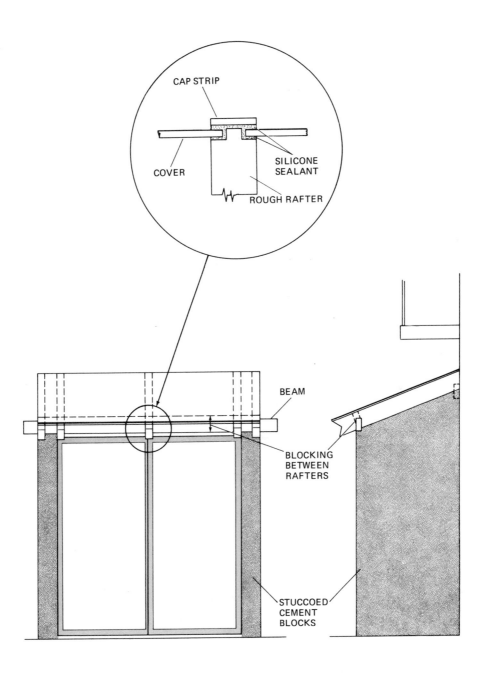

CAP STRIP

COVER

SILICONE SEALANT

ROUGH RAFTER

BEAM

BLOCKING BETWEEN RAFTERS

STUCCOED CEMENT BLOCKS

LEAN-TO GREENHOUSE

This custom lean-to greenhouse can be framed a couple of ways. National Greenhouse roof end bars (No. 124) could be used at the ends with gable bars (No. 117) in the center. Roof end bar (No. 124) could also be used at the top with some trimming. The construc-

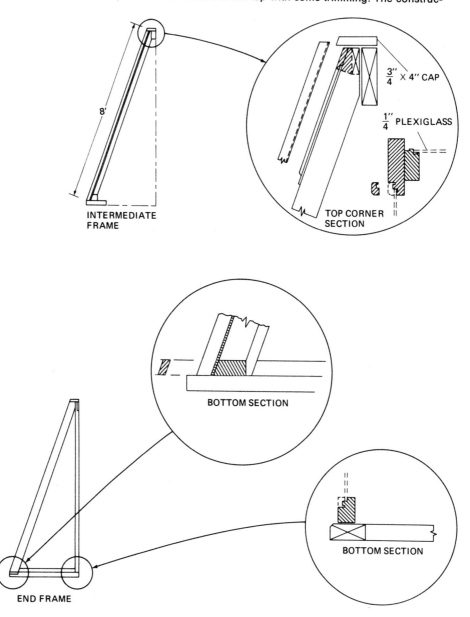

INTERMEDIATE FRAME

$\frac{3''}{4}$ × 4'' CAP

$\frac{1''}{4}$ PLEXIGLASS

TOP CORNER SECTION

BOTTOM SECTION

BOTTOM SECTION

END FRAME

8'

tion shown uses all 2-inch, construction heart redwood and involves some table-saw or router work to achieve the shapes. Wide-board lumber lets you select stock that is relatively free of knots. Quarter-inch Plexiglas is the best choice for the cover. All seams should be caulked with silicone.

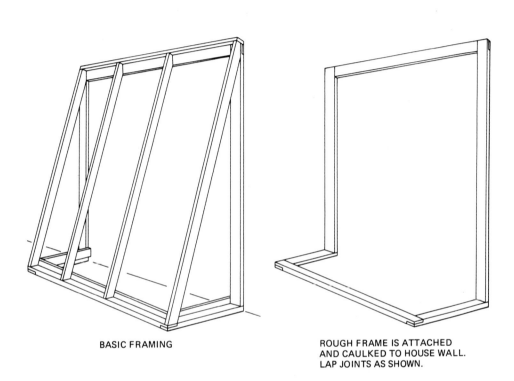

BASIC FRAMING

ROUGH FRAME IS ATTACHED AND CAULKED TO HOUSE WALL. LAP JOINTS AS SHOWN.

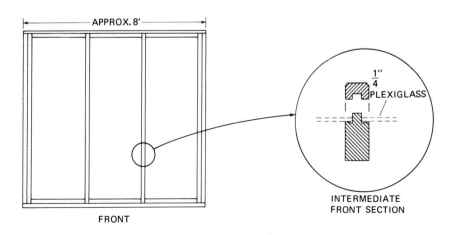

APPROX. 8'

$\frac{1}{4}''$ PLEXIGLASS

FRONT

INTERMEDIATE FRONT SECTION

A-FRAME GREENHOUSE

A-frame greenhouses are especially attractive and spacious-looking when the roof and the ends are covered in clear plastic or glass. A foundation of concrete will resist lateral forces of the roof bars.

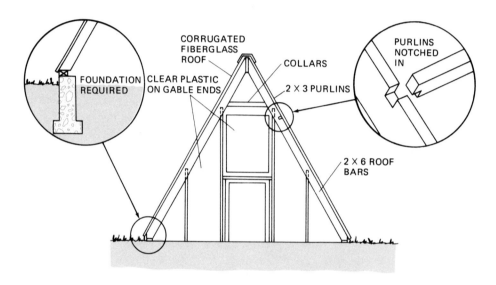

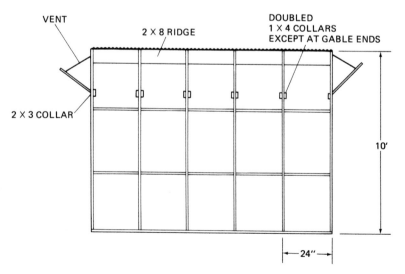

WINDOW GREENHOUSE

To build this, first cut the redwood frames and then cut and fit the Plexiglas. For a better seal, use thickened cement rather than solvent cement. Finish the redwood frames; then mount to the assembled Plexiglas. Caulk with silicone. Assemble the window over the seat board and add bottom trim.

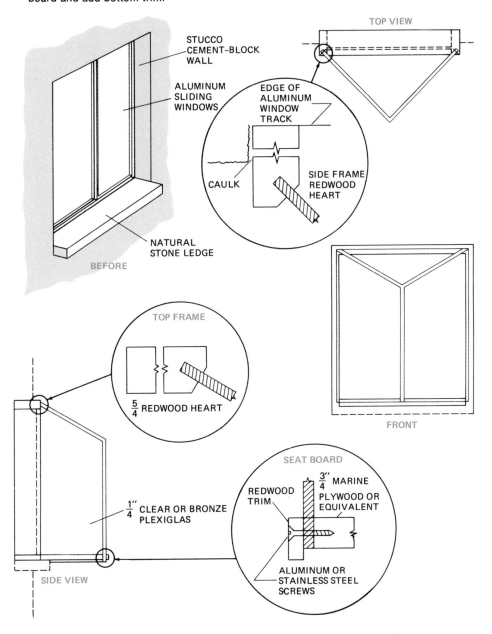

PIT GREENHOUSE

A pit greenhouse has low initial cost and low heating cost because the earth serves to insulate the walls. A south-facing hillside makes an excellent location. Excavate four feet; build walls of concrete blocks or of poured concrete on a poured footing. The entryway will probably have to be sunken. Design such a pit greenhouse around sashes that are available at your home supply center. Have glass on the south side only. Insulate all above-ground walls and the shingled portion of roof. In winter, cover the glass before sunset with insulating panels or blankets of some sort to conserve heat. Double glazing helps. A small electric heater or a wood stove will get you through most cold nights, and a small fan will promote air circulation.

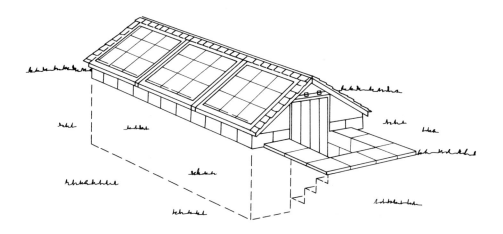

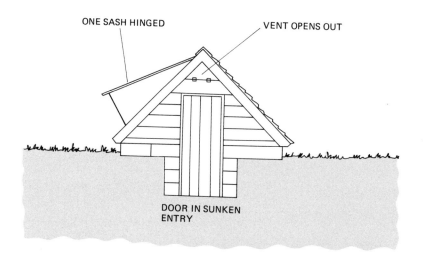

ONE SASH HINGED

VENT OPENS OUT

DOOR IN SUNKEN ENTRY

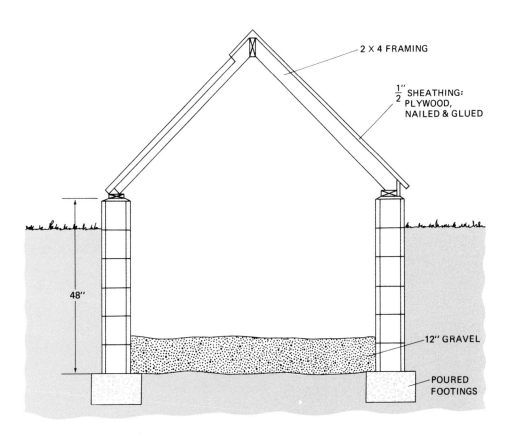

2 X 4 FRAMING

$\frac{1}{2}''$ SHEATHING:
PLYWOOD,
NAILED & GLUED

48''

12'' GRAVEL

POURED
FOOTINGS

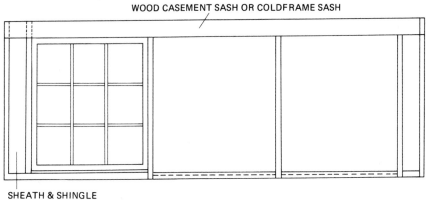

WOOD CASEMENT SASH OR COLDFRAME SASH

SHEATH & SHINGLE

REDWOOD BENCH

A redwood greenhouse bench can be constructed of common redwood lumber and fittings that can be bought in most hardware stores or home centers. Legs that are 36 inches give most people a comfortable working height, but for less waste of space in the greenhouse, particularly in a greenhouse that is not glass-to-ground, shorter legs are more practical. The bench can be constructed with a redwood-plank bottom, or with 1 × 2 turkey wire supported on 2 × 2 cleats. Assemble all wood parts with aluminum drive-nails. Then add mending angle reinforcement.

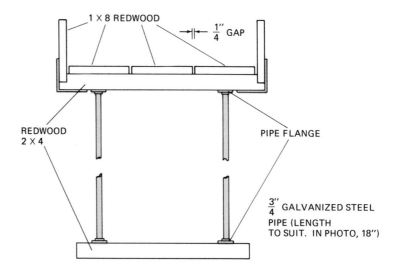

1 × 8 REDWOOD

$\frac{1}{4}''$ GAP

REDWOOD
2 × 4

PIPE FLANGE

$\frac{3}{4}''$ GALVANIZED STEEL
PIPE (LENGTH
TO SUIT. IN PHOTO, 18'')

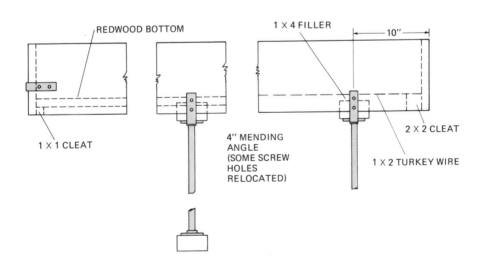

REDWOOD BOTTOM

1 X 4 FILLER

10"

1 X 1 CLEAT

4" MENDING ANGLE (SOME SCREW HOLES RELOCATED)

2 X 2 CLEAT

1 X 2 TURKEY WIRE

Constructing Foundations

All greenhouses require a foundation of some sort. A foundation serves two main purposes. First, it provides a level surface on which to erect the greenhouse frame. (Of course, you can adapt a greenhouse for bases that are not level, but most kit greenhouses are designed for level bases.) Second, a greenhouse foundation provides a means of attaching the greenhouse to the ground or to a platform so it doesn't shift around or blow away in a high wind. A foundation can also correct or prevent a drainage problem.

Foundations for greenhouses can take many forms, depending on greenhouse design, location on your property, requirements of the local code, and on whether the greenhouse is to be a permanent or temporary structure.

MASONRY WALL

A masonry wall foundation consists of a footing and a foundation wall. The footing supports the weight of the foundation wall, and the weight of the greenhouse. In general construction practice, the footing is twice as wide as the wall, and its height is equal to the thickness of the wall. To provide required support, the bottom of the footing must be flat and level, and must rest on undisturbed soil below the frost line.

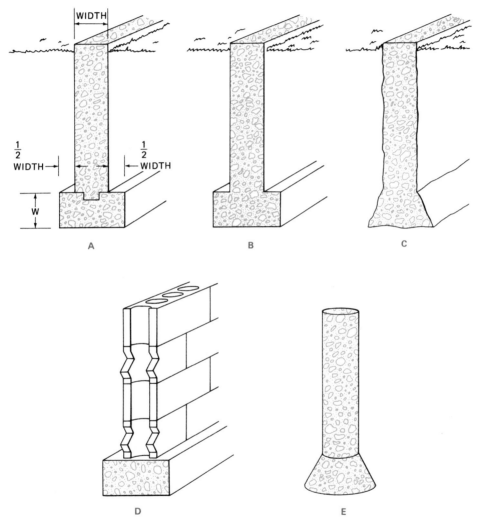

Here are the masonry foundations you can use for greenhouses. **In A**, footing and foundation were poured separately in forms. **In B**, footing and foundation were poured into a wooden form in one operation. **In C**, concrete was simply poured into a trench, with a leveling form used only at the top. **In D**, a concrete-block foundation was laid over a concrete footing. **In E**, a pier was made by pouring concrete into a hole dug with post-hole digger or earth auger.

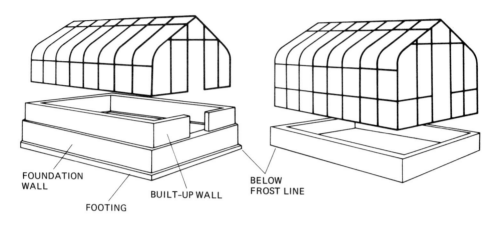

FOUNDATION
WALL

FOOTING

BUILT-UP WALL

BELOW
FROST LINE

FOUNDATION A

FOUNDATION B

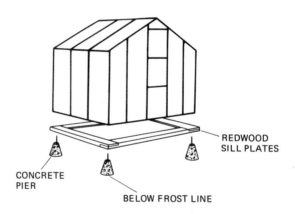

CONCRETE
PIER

BELOW FROST LINE

REDWOOD
SILL PLATES

FOUNDATION C

These are the basic types of foundations for glass greenhouses. **Foundation A** consists of a poured footing, a poured or cement-block foundation wall, and a built-up wall which can be poured concrete, cement block, brick, or stone masonry. **Foundation B** features a simple, poured trench-wall foundation. Designed for lightweight greenhouses, **Foundation C** consists of corner-lapped 2 × 6 redwood heart lumber attached to concrete earth anchors at the corners. Since a lightweight greenhouse is not likely to be damaged by frost heaving, local building codes may leave depth of the anchors up to you. Yet, if such a lightweight greenhouse is attached to a building, then the foundation must be adequate to prevent heaving that would otherwise cause the greenhouse to separate from the companion building.

The foundation wall may be brick, poured concrete, cement or cinder block, or mortared stone masonry. A masonry foundation combined with a proper footing placed below the frost line can prevent damage from heaving—the movement of the ground caused by freezing and thawing. A reinforced concrete foundation will resist earthquake shock.

POURED CONCRETE

Concrete is made by pouring, or casting, wet concrete into spaces enclosed by forms. The wet concrete hardens into the shape outlined by the forms, after which the forms are usually removed.

For greenhouse foundations, it's easiest to use the sides of the earth excavation to form the mold for the footing. Besides saving effort in digging an oversize trench, and then the effort and cost for making a wood form for the footing, a simple trench actually produces a more secure earth-compacted footing. If you are going to insulate the foundation wall (see Chapter 6), place a form on the inside of the trench, leaving room for insulation later placed between the concrete and the trench wall.

Exterior plywood is the best material for the forms. Lumber or board forms are more difficult to make. Besides, they tend to warp, and they are usually not as tight, smooth, and knot-free.

If the forms are not tight, wet concrete and water will leak out through the cracks. The forms must also be strong and well-braced to resist the high pressure exerted by the concrete and the tamping process. In addition to bracing outside the forms, wire ties connect opposite posts, passing through the forms and concrete. When the forms are removed, the wires are cut off close to the concrete surface.

REINFORCED CONCRETE

Concrete has strong compression resistance, but it is weak under tension; the reverse is true for slender steel rods. When steel rods are embedded in concrete each of the two materials makes up for the deficiency of the other, and the combination is known as reinforced concrete. The steel may consist of welded wire mesh, expanded metal mesh, or steel bars called *reinforcing bars*, or *rebars* for short. Rebars are usually employed in greenhouse foundations.

Before placing reinforcing steel in forms, all the forms should be oiled so they can be removed easily after the concrete has set. Oil spilled on reinforcing bars or mesh reduces the metal bond to the

This shows basic wooden form work used for a foundation poured onto a footing.

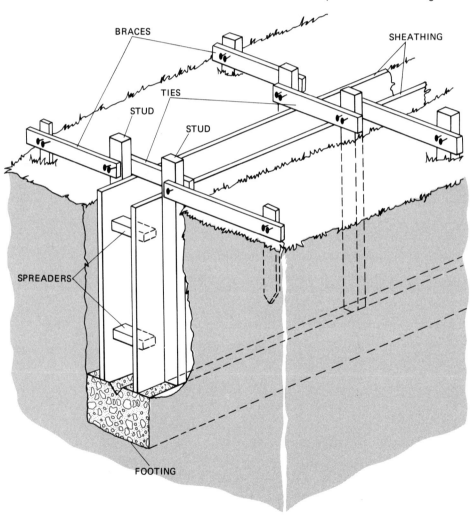

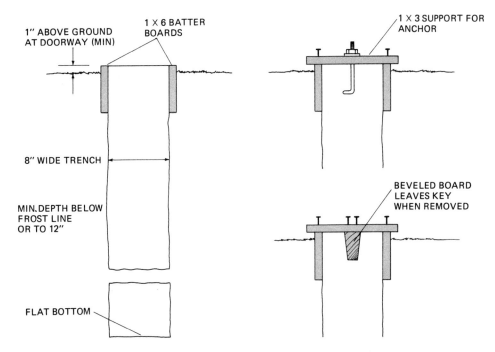

Before pouring a trench-wall foundation, you must construct leveled batter boards around the top edge. These batter boards also support hardware such as anchor bolts and beveled boards that leave keys in the concrete when removed.

concrete. A film of rust or mill scale on the steel does not have to be removed unless it shows signs of flaking.

Concrete will not attain its maximum possible strength, density, and uniformity unless it is properly placed in the forms. If concrete were simply poured into a form and left to harden, large numbers of air bubbles (called honeycomb) would be entrapped in the material, resulting in inadequate density. Surfaces would not be solid, and coarse aggregate and rock pockets would be exposed. Form spaces would not be completely filled, and the reinforcing steel would not be solidly bonded to the concrete.

To eliminate these and other defects, concrete must be consolidated by means of spading. In spading, the spade tool is shoved down along the inside surface of the form through the layer to be spaded and several inches into the layer previously placed. The tool is then worked back and forth several times until the air has been expelled from the concrete and until all the form spaces are well filled.

CURING CONCRETE

Concrete hardens as a result of the water's hydrating the cement. Freshly placed concrete contains more than enough water to hydrate the cement completely, but if the concrete is not protected against drying out, the water content, especially at the surface, will drop below that required for complete hydration.

Thus it's necessary to cure ordinary concrete by preventing surface evaporation of water during the period between beginning and final set. Concrete takes a beginning set in about one hour. Final set is not completed for about seven days. All you have to do is keep the surface from drying out. This can be done by spraying with water and covering the concrete surface with newspaper or polyfilm.

Concrete should be protected from direct sunlight for at least the first three days of the seven-day curing period. Wood forms should be loosened and removed when the concrete is hard enough to hold its own weight safely. Prior to this the space between the forms and the concrete should be flooded with water at frequent intervals. But if you want to reuse the forms, be sure to strip drying concrete off them as soon as possible. If you wait too long, the concrete will become stubbornly bonded to the wood.

TRENCH WALL FOUNDATION

Since greenhouses—even the heaviest—weigh little compared to the weight of a two-story house, some greenhouse manufacturers recommend a simple, poured trench wall with no special footing, and extending below the frost line. A trench wall foundation incorporates the footing and wall in a single poured concrete mass, usually poured straight into the narrow trench without forms except near the exposed top. Your local building code may not allow this light-load distinction, however, and may require a heavier foundation, complete with footing.

A trench wall foundation is the easiest masonry foundation to construct. You can rent a walk-behind trencher for about $60 a day. This trencher will claw a trench 6 to 8 inches wide (depending on vibration) and 24 inches deep. With such a trencher, it shouldn't take more than an hour and a half to dig a trench for a whole greenhouse foundation. The trench itself becomes the form for the concrete. A trench wall foundation, however, cannot be insulated.

Note: Before placing any concrete or concrete block foundations, determine the exact dimensions required for your greenhouse and the required anchoring provisions, particularly if it is a kit. When you order a kit greenhouse requiring a masonry wall foundation the manufacturer will send the foundation plans ahead.

CONCRETE BLOCK FOUNDATION

A foundation made of concrete block can be used instead of one of poured concrete. You'll still have to pour a concrete footing however. The blocks should be laid in a running bond. That is, each 16-inch-long block overlaps two blocks in the course below half way. A running bond foundation wall has greater lateral strength than one made with any other bond, and the cells of the blocks line up vertically so rebars can be easily inserted and grouted (packed with mortar).

Plan your foundation so you don't have to cut blocks, which is not as easy as cutting bricks. Before you dig, lay out a course of blocks on the ground, determine how your greenhouse will fit, then excavate accordingly. Although you can start laying blocks on the poured footing while it is still soft, giving you a stronger foundation, you'll find it easier to wait until the footing is dry and hardened.

On a dry footing lay the first course dry to locate the corner blocks. Then mark the positions of all of the blocks, and remove them. Clean off and wet down the footing. Then trowel on a ½-inch deep bed of mortar the size of a corner block and set the block, making sure it is level in both length and width. Press the block down to make a ⅜-inch thick joint; then do the next corner block and all the blocks in between, buttering the vertical edges for bonds to adjacent blocks. Now repeat for the other sides of the foundation. When mortaring the rest of the courses, butter only the vertical and horizontal edges along each block's inside and outside.

Vertical reinforcing rods, when required, are set into the footing and positioned to come up through the cells in the concrete blocks. After the wall is laid, fill the cells containing rebars with a mixture of gravel and grout. Horizontal reinforcing can be rods laid in the mortar between the courses.

After setting the anchors for the greenhouse structure, seal the top of the wall to keep out moisture by filling the cells of the top course with grout. You can prevent the grout from disappearing into the wall by stuffing in balls of newspaper first.

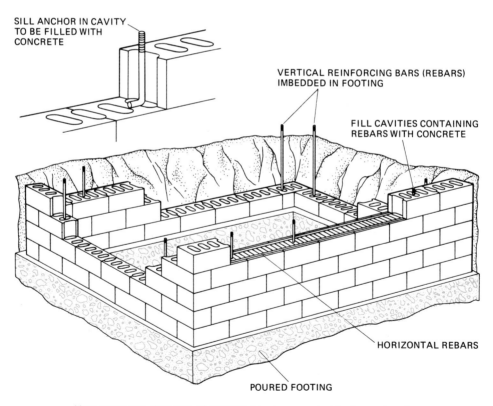

SILL ANCHOR IN CAVITY TO BE FILLED WITH CONCRETE

VERTICAL REINFORCING BARS (REBARS) IMBEDDED IN FOOTING

FILL CAVITIES CONTAINING REBARS WITH CONCRETE

HORIZONTAL REBARS

POURED FOOTING

Here are some elements of running-bond, concrete-block construction.

CONCRETE & CINDER BLOCKS

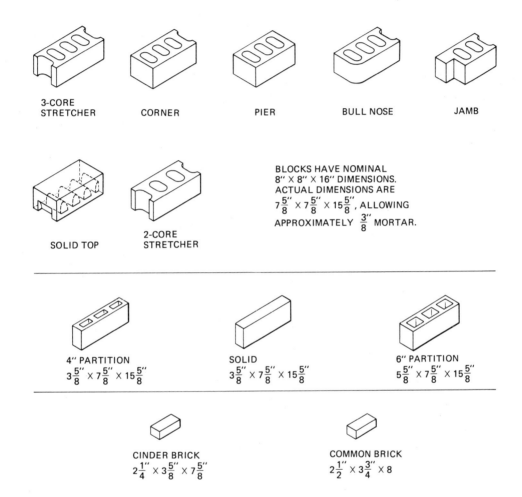

3-CORE
STRETCHER

CORNER

PIER

BULL NOSE

JAMB

SOLID TOP

2-CORE
STRETCHER

BLOCKS HAVE NOMINAL
8″ X 8″ X 16″ DIMENSIONS.
ACTUAL DIMENSIONS ARE
$7\frac{5}{8}$″ X $7\frac{5}{8}$″ X $15\frac{5}{8}$″, ALLOWING
APPROXIMATELY $\frac{3}{8}$″ MORTAR.

4″ PARTITION
$3\frac{5}{8}$″ X $7\frac{5}{8}$″ X $15\frac{5}{8}$″

SOLID
$3\frac{5}{8}$″ X $7\frac{5}{8}$″ X $15\frac{5}{8}$″

6″ PARTITION
$5\frac{5}{8}$″ X $7\frac{5}{8}$″ X $15\frac{5}{8}$″

CINDER BRICK
$2\frac{1}{4}$″ X $3\frac{5}{8}$″ X $7\frac{5}{8}$″

COMMON BRICK
$2\frac{1}{2}$″ X $3\frac{3}{4}$″ X 8

PIER FOUNDATION

Except for the heaviest glass greenhouses, a combination of piers—
concrete columns or wood posts—and sill plates can form a satisfactory
foundation, and involve a lot less effort and cost. Cylindrical columns
of reinforced concrete can be poured directly into holes dug with a

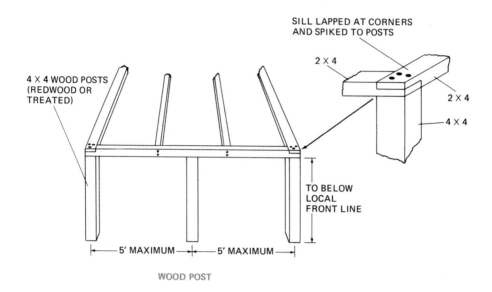

4 X 4 WOOD POSTS
(REDWOOD OR
TREATED)

SILL LAPPED AT CORNERS
AND SPIKED TO POSTS

2 X 4

2 X 4

4 X 4

TO BELOW
LOCAL
FRONT LINE

5' MAXIMUM   5' MAXIMUM

WOOD POST

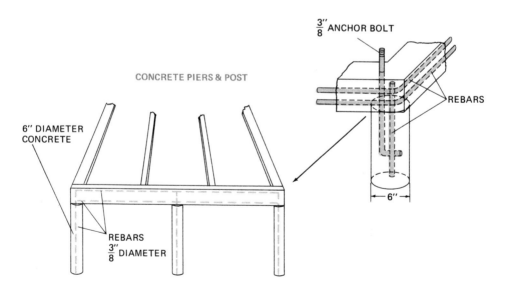

CONCRETE PIERS & POST

$\frac{3}{8}''$ ANCHOR BOLT

REBARS

6'' DIAMETER
CONCRETE

REBARS
$\frac{3}{8}''$ DIAMETER

6''

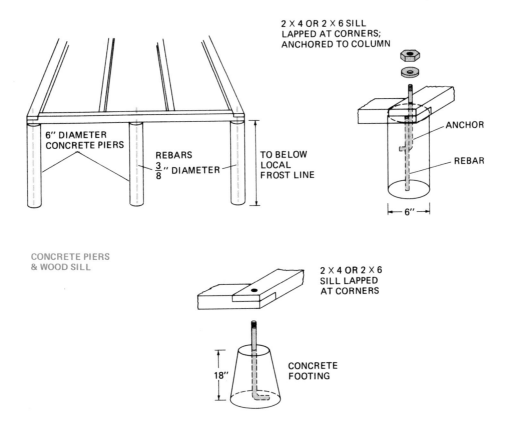

2 X 4 OR 2 X 6 SILL
LAPPED AT CORNERS;
ANCHORED TO COLUMN

6" DIAMETER
CONCRETE PIERS

REBARS
$\frac{3}{8}$" DIAMETER

TO BELOW
LOCAL
FROST LINE

ANCHOR

REBAR

6"

CONCRETE PIERS
& WOOD SILL

2 X 4 OR 2 X 6
SILL LAPPED
AT CORNERS

18"

CONCRETE
FOOTING

post-hole digger or earth auger. Holes should be dug to below the frost line. Before pouring concrete, suspend reinforcing bars in place in the holes and have sill anchors ready. Locate the piers at the corners of the greenhouse and not more than five feet apart along the sides and ends.

Sill plates for greenhouses are usually redwood 2 × 4s or 2 × 6s, single or doubled, lapped at the corners. Reinforced concrete sills are also used. Redwood 4 × 4s, or 6 × 6s, or treated wood posts can be substituted for the reinforced concrete piers.

SILL PLATE

Lightweight greenhouses, temporarily installed greenhouses, and even light glass-covered free-standing greenhouses can be erected on foundations consisting of 2 × 4 or 2 × 6 single sill plates, lapped at the corners. The sill plates must be anchored to the ground by some means. Otherwise, in below-freezing weather, they probably will move when the ground heaves. I have had a glass-covered greenhouse set up this way for three years now in the cold Northeast without any structural problems.

There is a variety of schemes for anchoring the sill plate. Concrete anchors poured into holes in the ground flared at the bottom are the best. Suspend an anchor bolt in the hole before you pour. The bolt must be long enough to extend one inch through the sill plates, and must be located so as not to interfere with the structure of the greenhouse.

Some kit manufacturers recommend using pipes or 2 × 4 stakes as anchors for the sill plate. But these anchors have limitations. First, they

Anchor bolts, like these, and rebars used in foundations should be ⅜-inch in diameter.

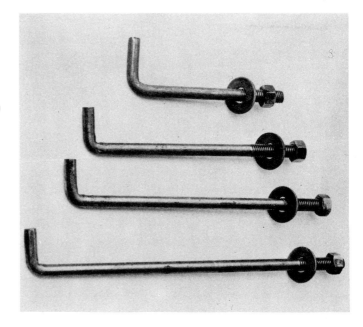

can't be driven in all types of soil. Second, it is almost impossible to drive them in so they are located just where you want them. Third, there's no really good way to attach the sill plates to them.

Small, temporarily installed greenhouses with film covers often have no sill plate except the bottom framing member. This is sometimes anchored with a hooked metal stake. A better way to anchor a small greenhouse is to use earth anchors that resemble giant corkscrews. These earth anchors can be installed by hand-twisting them into the ground. Lines are passed over the top of the greenhouse and secured to the eyes.

5

How to Achieve Climate Control

Whether you build from a greenhouse kit or from your own design, when you've erected the shell you're still not done. You have to provide the climate required for your plants. This can be a simulation of anything from northern summer conditions to those of a tropical jungle. To achieve the desired climate, you must control temperature, humidity, ventilation, and hours of sunlight. Fortunately most, if not all, of the climate control can be done automatically.

HEATING CONSIDERATIONS

The cost of heating a greenhouse depends almost entirely on the following factors:

- What you want to grow
- Where you live
- Greenhouse size and construction
- Location of the greenhouse on your property
- Choice of heating system
- Insulation and other heat conservation efforts

HEATING NEEDS

Your choice of heating method is an important one, and it should be a factor in selecting the site on your property. Site determines your heating options and the amount of heating required. If your greenhouse is attached to your home, or close by it, you might be able to extend your home steam, hot-water or hot-air heating system to handle all or most of the greenhouse heating load. This is the most economical way to heat a greenhouse, except possibly by installing a solar-heating collection and storage system.

Old-fashioned advice was to first check local gas, oil, and electric rates before making a decision on the kind of heater to use. But shortages of oil and natural gas have changed greenhouse heating guidelines. If you plan to invest a lot of time, effort, and money in a greenhouse and its plants, you want to be reasonably sure the heat won't be cut off, and that you have a backup means of heating just to be safe. As a minimum, the backup heater should be able to maintain enough heat for plant survival if not for growth.

Plants have different temperature requirements, and a single plant can have different requirements at different times in its growing cycle. Most seeds require a temperature of 65° to 70° F to germinate, but many seedlings do not require the same high temperature to continue growing. Greenhouses are customarily operated in one of three minimum night temperature ranges:

Cool greenhouse	45—50°F
Moderate greenhouse	55—60°F
Warm greenhouse	65—70°F

In a cool greenhouse you can grow asparagus, beets, carrots, lettuce, radishes, parsley and swiss chard. In a moderate greenhouse, you can grow broccoli, cauliflower, and spinach. Tomatoes require a warm greenhouse. And there's a multitude of flowers and decorative plants for all of the temperatures, as shown in table on the next page (Greenhouse Plants).

In the colder areas of the country, the cost of maintaining a cool night temperature in a greenhouse is about half the cost of heating for warm greenhouse temperatures.

It is possible to operate the end of the greenhouse nearest the heater as a warm greenhouse and the rest as a cool or moderate

GREENHOUSE PLANTS (Minimum Night Temperature Ranges)

Cool Greenhouses 45–50° F

Acacia
Agapanthus (Lily of the Nile)
Ageratum (Floss Flower)
Anemona (Windflower)
Aucuba (Gold Dust Tree)
Calceolaria (Lady's Pocketbook)
Calendula (Pot Marigold)
Camellias
Campanula (Bellflower or Falling Star)
Chrysanthemums
Dianthus (Carnation)
Eschscholzia (California Poppy)
Freesia

Gypsophila (Baby's Breath)
Lathyrus (Sweet Pea)
Lobellia
Lobularia (Sweet Alyssum)
Myosotis (Forget-me-not)
Pansy
Plumbago (Leadwort)
Primula (Primrose)
Ranunculus
Saxifraga (Strawberry Begonia)
Schizanthus (Butterfly Flower)
Tritonia
Tropaeolum (Nasturtium)

Moderate Greenhouses 55–60° F

Achimenes (Magic Flower)
Amaryllis
Antirrhinum (Snapdragon)
Azalea
Begonias
Beloperone (Shrimp Plant)
Bougainvillea
Bouvardia
Browallia
Coleus
Cyclamen
Daffodil
Dahlias
Euphorbia (Poinsettia)
Ferns

Fuchsia (Lady's Ear Drops)
Gladiolus
Hyacinth
Iris
Kalanchoe
Lilium (Lily)
Manettia (Firecracker Vine)
Passiflora (Passion Flower)
Pelargonium (Geranium)
Peperomia
Philodendron
Rosa (Roses)
Tulips
Zantedeschia (Calla Lily)
Zygocactus (Christmas Cactus)

Warm Greenhouses 65–70° F

Aeschynanthus (Lipstick Plant)
Asparagus Plumosus (Asparagus Fern)
Bromeliads
Cacti and Succulents
Caladium
Clerodendrum (Glory Bower)
Eucharis (Amazon Lily)

Euphorbia (Crown of Thorns)
Gardenia
Gloriosa (Glory Lily)
Orchids
Saintpaulia (African Violet)
Sinninglia (Gloxinia)
Tradescantia (Spiderwort)

greenhouse by utilizing the temperature gradient between the heater and the far end, or by judicious partitioning. You can maintain higher temperatures in small areas for germinating seeds, for seedlings, or for force growing by using flexible electric heater cables in the seed beds plus tents over the beds—in effect creating little greenhouses within the greenhouse.

The severity of your winters, measured in the number of degrees the daily average outdoors temperature is below 65°F (degree days), will determine the annual heating costs. The maximum capacity of the heater or heating system should be based on the lowest outdoor temperature expected during the heating season. Heat output capacity is measured in British thermal units per hour (Btuh).

In the next chapter, you'll find a table that indicates lowest expected temperatures and heating season degree days for many locations throughout the U.S. The table below shows estimated heating requirements in Btuh for a 55° F greenhouse in Philadelphia.

HEATING LOAD FOR TYPICAL GREENHOUSES

Type	Length	Width	Area Floor Sq. Ft.	BTUH Heat Load*	Electric Heater Size
Small Film Covered Gable	8' 0"	6' 0"	48	13150	3.5 kw
Medium FRP Covered Lean-to	16' 0"	7' 0"	112	19410	5.2 kw
Medium FRP Covered Gable	16' 5"	6' 10"	112	25410	6.8 kw
Large Glass Covered Gable	19' 0"	9' 10"	187	35470	9.5 kw
Large Glass Covered Gable	21' 2"	14' 0"	296	47250	12.7 kw

*This heat load is for a 55°F warm greenhouse located in surburban Philadelphia. Winter design temperature allows for a coldest day of 11°F and 4000 degree days. See the next chapter for the method of calculating needs. Heating calculations are for average construction and single-layer covers.

HEATING EQUIPMENT

Small greenhouses are usually heated by portable electric space heaters. Types with circulating fans are preferred over radiant heaters because their heat is spread more evenly, and the air is circulated better. Thermostat control is better than simple ON/OFF or two-speed control. But for best control, use a separate thermostat that senses greenhouse temperature rather than just heater temperature. The ther-

Many small greenhouses can be economically and conveniently heated with one or more portable electric heaters. Heat outputs range from 5,600 to 19,000 Btuh. High ratings require 208- or 240-volt power rather than 120 volts. The leftmost heater shown is a Chromalux. The other two are Titans.

This electric unit heater provides convenient total heating for a medium-sized greenhouse, or supplemental heating for a larger greenhouse. No venting is required. Heat outputs range from 17,000 to 85,000 Btuh (power input, 5 kw to 25 kw). Unit heaters run off 208- or 240-volt supplies. (*Courtesy Modine Manufacturing Co.*)

mostat should be located in a position that you have determined optimal for your plants.

In a small greenhouse the electric heater is usually placed under the bench at one end and angled to blow heat the length of the greenhouse. To supply the heater, you will need a separate 115- or 230-volt circuit run from your house power panel. (The largest electric heater you can run off a single 115-volt, 20-A branch circuit will give you about 5,000 Btuh.) Several small electric space heaters will warm the greenhouse more evenly than one large unit.

Portable electric heaters come in a variety of ratings. There is no need to look for one specifically made for greenhouses. But be sure to select a heater with approval of the Underwriters' Laboratories (UL) and a grounded plug.

Outside-venting Gas Heaters

For small and intermediate size greenhouses, outside-venting gas heaters are the most practical type. They are made in three ratings: 15,000, 26,000, and 33,000 Btuh. In them, the combustion chamber is sealed away from inside the greenhouse so that no oxygen is taken

TYPICAL ELECTRIC UTILITY HEATERS

Type	Voltage	Watts	Btu	Required Circuit Capacity	Approximate Cost
Single Range	120	1650	6,000	20 A	$ 31
	120	1500	5,600	20 A	163*
	240	3000	10,200	20 A	69
	240	4000	13,700	25 A	70
	240	5600	19.000	30 A	150**
Dual Range	120	1300/1500	4,400/5,100	20 A	28
	240	3600/4800	12,200/16,400	20 A	195

Note: * With Calrod element
       ** With bracket mount

This thermostatically controlled Dyna-vent gas heater is mounted in a green-house much the way a window aircon-ditioner is. Because the heating unit is sealed from inside and vented outside, the heater is safe for all plants. Heat outputs for different models range from 15,000 to 34,000 Btuh. The Dyna-vent is available from several green-house suppliers.

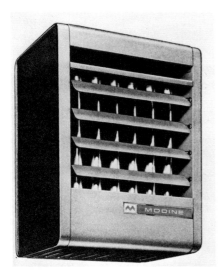

Gas-fired unit heaters come in propel-lor and blower types for overhead mounting. Heat outputs range from 24,000 to 160,000 Btuh. These heaters require only venting and connection to gas and electric services. Operation can be manual or automatic, with the thermostat located away from heater. (*Courtesy Modine Manufacturing Co.*)

from the growing area. Thus, they use outside combustion air, and the products of combustion are safely expelled by fan outdoors. The heaters are American Gas Association approved and have safety pilots or electric spark ignition and high-limit gas shutoff switches. Masonry chimneys or metal flues are not required. (Wood stoves are not practical mainly because the clearances required would waste a lot of potential growing space, and wood stoves require periodic reloading around the clock.)

Gas heaters are controlled by separate thermostats, which are supplied with the heaters. For greenhouse heating requirements above the capacity of the heater, two units may be more economical to operate than one large gas or oil furnace. If you do use two together, operate each from a separate thermostat and set one thermostat 5°F higher. That way one heater takes all of the load in moderate weather and the second heater comes on only to handle the heating load beyond the capacity of the other.

Gas heaters with outside vents are installed in masonry or in solid greenhouse walls, or else in place of a glass light in glass-to-ground designs. When heaters replace glass, they are mounted in an aluminum panel cut to fit the space.

NO-VENT GAS HEATERS

Btuh rating	Wall opening (inches)	Weight (pounds)	Approx. price with thermostat
15,000	15 × 15″	85	$290
24,000	20 × 17″	95	312
26,000	20 × 17″	110	330
23,000	20 × 17″	115	360

Gas- and Oil-fired Furnaces

Gas- and oil-fired furnaces supply 64,000 to 140,000 Btuh heating loads. These discharge warm air near floor level for heating the under bench areas. The furnaces can be used with ducts if necessary for heat distribution in long greenhouses. A. G. A. approved models are supplied with necessary controls. For these, local code will require masonry chimneys or metal flues. Gas-fired types can be run on either natural or bottled gas, but not manufactured gas because it will kill plants.

The furnaces are made in upright and horizontal styles. The horizontal style can be located either under a bench or hung overhead.

If you use a vented, gas- or oil-fired heater, be sure to provide

Here are oil/gas furnace options for greenhouses. The horizontal style can be installed under a bench, or attached to the roof of a large greenhouse. The vertical style takes up little floor space. Both oil- and gas-fired furnaces require stacks. Both types are available in four maximum capacities ranging from 84,000 to 140,000 Btuh.

enough fresh air for combustion. A small hole about 3 inches in diameter through the wall is usually adequate. When automatic greenhouse ventilation is used, the heater must be linked with a thermostat that does not allow the heater to operate with the vents open. If a heater operates with greenhouse vents open, a downdraft in the chimney or heater vent could cause noxious fumes to accumulate inside the greenhouse. This would be harmful to plants and people. To prevent this from occurring, set the ventilating thermostat 10 degrees higher than the heating thermostat.

Unit Heaters

These compact space heaters are mounted overhead in the greenhouse. Depending on type, they are heated by steam, hot water, gas, oil, or electricity. They use either propellor or blower fans and can deliver air downward or horizontally. For medium and large home greenhouses, unit heaters provide flexibility in heating, and also can be used to provide air circulation.

GAS-FIRED AND OIL-FIRED WARM AIR FURNACES (J. A. NEARING CO.)

VERTICAL

Gas-Fired					Oil-Fired				
Btu Output	Width	Depth	Height	Price	Btu Output	Width	Depth	Height	Price
64,000	16″	28″	55″	$260	84,000	20″	33 ½″	65″	$474
84,000	20″	28″	55″	280	95,200	20″	33 ½″	65″	534
96,000	22 ½″	28″	64″	375	112,000	24″	33 ½″	65″	580
112,000	22 ½″	28″	64″	480	140,000	27″	33 ½″	68″	710

HORIZONTAL

Gas-Fired					Oil-Fired				
Btu Output	Width	Depth	Height	Price	Btu Output	Width	Depth	Height	Price
64,000	50″	16″	24″	$280	84,000	68″	23″	20″	$476
80,000	50″	18″	24″	310	95,200	68″	23″	30″	546
96,000	58″	25″	24″	405	112,000	68″	23″	24″	582
112,000	58″	25″	24″	450	140,000	68″	23″	24″	720

Gas- and oil-fired unit heaters require metal stacks or masonry chimneys. Gas-fired units have safety pilot lights. Hot water and steam units are suitable only for large greenhouses with their own furnaces or boilers. Electric unit heaters can be used for standby or supplemental heating, or for total heating where other fuel supplies are not available.

A major benefit of unit heaters in greenhouses is positive circulation of air, and this discourages plant mold and fungal diseases. Delivery of heat from above the plants slows evaporation and lowers moisture accumulation on plants. This not only promotes healthier plant growth, but reduces the need for watering. But you should avoid blasting heated air directly on the plants because it can damage them.

A room thermostat can control the cycling of fans and the heater itself. With fans in continous operation, air is constantly circulated. This will minimize the possibility of cold spots by discouraging accumulation of water droplets on crops and by circulating air so that the thermostat can more rapidly detect a change in temperature. Plus, the unit can operate as an air-circulating fan in warm weather.

Each unit heater should have its own thermostat. No more than two unit heaters should ever be controlled by a single thermostat. When each unit heater is controlled by its own thermostat with continuous fan operation, temperature is monitored at all times in the zone served by that heater. This results in more uniform temperatures throughout the greenhouse. When you want to hold different temperature levels in adjacent areas of the greenhouse, the use of one thermostat per heater is a must.

Heaters with propellor fans cannot be used with air discharge nozzles, polytubes, or ductwork. A heater with a fan has a fixed air delivery and air temperature rise (difference between inlet and discharge air temperatures), usually in the 55 to 65°F range. It is the most widely used unit heater because of its versatility, low cost, and compact size.

Heaters with blower fans are designed for use with ductwork or discharge air nozzles. Their operation is quieter than propellor types and can deliver the same amount of air. These units have a variable pitch motor sheave which permits adjustment of blower rpm and air volume to accommodate a range of air temperature rises from 55° to 85°F.

Unit heaters with horizontal delivery are used for low greenhouses. Horizontally positioned louvers attached to the air-discharge opening can be adjusted to lengthen or shorten heat throw. Adjustable vertical louvers can be used in combination with horizontal louvers to permit complete control of the direction of heated air.

Vertical units are better for heating areas with high roofs and where misting systems and other obstructions dictate high mounting of heating equipment. And louvers can be attached to provide a variety of heat-throw patterns.

Gas- and oil-fired hot water boilers can also be used economically for heating large home greenhouses. These circulate hot water around the perimeter of the greenhouse in aluminum-finned radiator pipes. This type of heating provides the most even heat distribution, but does not provide good air circulation. Furnaces come complete with boiler, all necessary controls, circulating pump, and thermostat.

When selecting components for your greenhouse heating system, allow some reserve capacity for matching the heater to the demands of an unusually cold winter. A moderately oversize heater will not cost any more to operate than one that exactly matches the load.

GAS-FIRED AND ELECTRIC UNIT HEATERS* (BY MODINE)

Gas Space Heaters

Model Number	Btu Output	Width	Depth	Height	Price**
PA30A	24,000	12 $\frac{7}{8}$"	14 $\frac{3}{4}$"	24 $\frac{1}{4}$"	$202
PA50A	40,000	17 $\frac{1}{4}$"	14 $\frac{3}{4}$"	24 $\frac{1}{4}$"	211
PA75A	60,000	17 $\frac{1}{4}$"	20"	28 $\frac{3}{8}$"	223
PA105A	84,000	19 $\frac{1}{4}$"	20"	28 $\frac{3}{8}$"	250
PA130A	104,000	21"	22"	35 $\frac{1}{4}$"	303
PA150A	120,000	21"	22"	35 $\frac{1}{4}$"	340
PA175A	140,000	23 $\frac{1}{2}$"	22"	35 $\frac{1}{4}$"	386
PA200A	160,000	25 $\frac{5}{8}$"	25"	40 $\frac{1}{4}$"	404

* Prices do not not include thermostats.
** Price for LP gas. Prices for natural gas are about $16 lower.

Electric Space Heaters (208 or 240 Volts)

Model Number	Btu Output	Kw	Width	Depth	Height	Phase	Price
HE50	17,065	5	17 $\frac{1}{4}$"	14 $\frac{3}{4}$"	24 $\frac{1}{4}$"	1	$164
HE75	25,597	7.5	17 $\frac{1}{4}$"	14 $\frac{3}{4}$"	24 $\frac{1}{4}$"	1	208
HE100	34,130	10	17 $\frac{1}{4}$"	14 $\frac{3}{4}$"	24 $\frac{1}{4}$"	1	224
HE125	42,660	12.5	17 $\frac{1}{4}$"	14 $\frac{3}{4}$"	24 $\frac{1}{4}$"	3	264
HE150	51,195	15	17 $\frac{1}{4}$"	14 $\frac{3}{4}$"	24 $\frac{1}{4}$"	3	357
HE200	68,260	20	19 $\frac{1}{4}$"	20"	28 $\frac{3}{4}$"	3	555
HE250	85,325	25	19 $\frac{1}{4}$"	20"	28 $\frac{3}{4}$"	3	638

TAPPING HOME CENTRAL HEATING

Sometimes, the least expensive way to heat a greenhouse is to tap into your home heating system—whether it be steam, hot water, or warm air. This can only be done if the greenhouse is attached to, or close by, your house. Often, with an attached greenhouse and a direct doorway access to your house, the heat flow through the open door is enough, especially when helped with a small fan.

Not all home heating systems can be extended, however. Be sure your furnace has the reserve heating capacity to take on the additional load which will reduce the heat available for your house. In extending the system to the greenhouse you are essentially adding a second (or third) heat zone complete with independent thermostatic control. Low-pressure steam boiler systems with forced circulation can usually be extended with little difficulty, but low-temperature hot water systems with gravity flow generally cannot be. Both are jobs for a plumber.

If your greenhouse is not attached to your home, the heating pipes should be buried in a trench below the frost line and well insulated to keep heat loss to a minimum. A house warm-air system cannot be extended too successfully to a separate greenhouse because of excessive heat losses through the ductwork.

VENTILATION

Ventilation is as important to greenhouse plants as winter heat. Greenhouse ventilation does four things:

1. It limits temperature rise in a greenhouse by allowing hot air to escape and be replaced with cooler outside air.
2. It controls humidity, which is caused by plant transpiration.
3. It replenishes the carbon dioxide in the greenhouse air. Carbon dioxide is necessary, along with light, for photosynthesis.
4. It helps control plant diseases and pests. Needless to say, vents should be screened to keep out insects and larger pests.

There are three means of ventilating a greenhouse: by natural convection, by exhaust fan, and by fan jet. Exhaust fan ventilation is used in all sizes and types of greenhouses. Natural convection ventilation is possible only in greenhouses with roof vents. Fan jet ventilation is practical only in very large greenhouses.

Vents and fans can be manually operated, but manual operation presents a drawback: Someone has to be on duty all day to open and close vents and switch fans on and off depending on the sunlight, the outside temperature and weather conditions, and the temperature inside the greenhouse. Fortunately there is a variety of automatic controls to take the drudgery and uncertainty out of greenhouse ventilation.

A thermostatically controlled exhaust fan system provides effective greenhouse ventilation. The basic system consists of an exhaust

With natural greenhouse ventilation, hot air is removed by means of a ridge-mounted vent sash. Cool outside air is admitted either by a low vent sash, as shown, or through an open or screened door. (*Courtesy British Aluminum Co., Ltd.*)

Here an electric motor moves levers that open vents extending the length of the greenhouse. Control is by means of a thermostat that can be set to open vents when greenhouse air is too hot and close them when the air is sufficiently cool. (*Courtesy Texas Greenhouse Co.*)

Here an exhaust fan on the rear wall and a motorized jalousie inlet vent are both controlled by a thermostat mounted on the right side wall (visible through the door). The thermostat is protected from sun-heating by a sun shade. This greenhouse is an English-style Sunlyt Even Span 6. (*Courtesy Lord & Burnham*)

fan mounted in one end of the greenhouse and an air inlet with either fixed louvers or shutters at the other end. A thermostat senses greenhouse temperature and turns the fan on when the temperature rises to a preset level. In small systems, the partial vacuum created inside the greenhouse opens the balanced inlet shutters. In larger systems, the inlet shutters are opened and closed by an electric motor. Motorized inlet shutters won't flap in the wind. For best results, the exhaust fan should be on the lee end of the greenhouse.

A two-speed fan is preferable to a single-speed fan. You will need ventilation in the winter, but you don't want any arctic blasts being drawn in. At the fan's high speed setting, you want it to be able to change the air every $1\frac{1}{2}$ to 2 minutes. A sidewall attic-type fan may be sufficient.

Although mounting the fan and inlet on opposite ends of a greenhouse is best, this is not always possible. Wherever fans are located, they should be as far apart as possible for most even ventilation.

Natural convection ventilation is provided by roof vents. The vents are square or rectangular sashes hinged to the ridge. In small greenhouses with manually operated ventilation, the vents are propped open with push rods. In larger greenhouses with vent sashes spaced all along the ridge, the sashes are opened and closed simultaneously by rack and pinion lifters operated by a pipe shaft that extends the length of the greenhouse. The pipe shaft is controlled by either a handcrank or a control wheel.

Fresh air is provided by jalousie sashes located below bench level in the sides or ends of the greenhouse. The jalousie units are usually operated individually. In a small greenhouse, a screened section of the door can function as one of the inlet vents.

There are several systems available for automatic control of roof vents. But whatever method is used, it is important that the vents open and close slowly to insure slow temperature changes. All systems are controlled by a thermostat which is set to the desired greenhouse temperature. When the inside temperature rises above this point, the vents open. When the temperature drops to a preset level, the vents are closed. Manual backup is provided in the event of a power failure. Exhaust fans provide ventilation in glass greenhouses when the temperature gets beyond the control of natural ventilation.

Jet fan heating and ventilating systems, such as Modine's Flora-Guard, are available for use in large commercial greenhouses. The

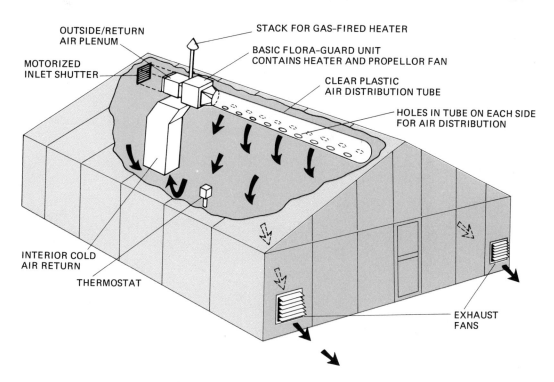

This is the Flora-Guard heating and ventilating system by Modine. While the smallest system made by Modine is too large for almost any home greenhouse, this drawing shows principles that can be applied in smaller setups.

Flora-Guard system consists of a gas-fired, steam or hot water heating and ventilating unit, an outside-return air enclosure, a polyethylene air-distribution tube, a motorized outside air inlet shutter, an exhaust fan, and heating and ventilating thermostats. The 24-inch diameter clear polyethylene tube is perforated with 3-inch holes punched at intervals along its length to evenly distribute air to the greenhouse. Tube length, hole size and location, and tube diameter are matched to the performance of the heating and ventilating unit.

With such a system, air circulation and temperature are continuous and uniform, and the system responds quickly to changes in thermal load. When the greenhouse becomes cool, the heating thermostat turns on the heating unit and, then, turns it off when the temperature has been brought up. When the temperature in the greenhouse builds up because of the sun's heat, the ventilating thermostat opens the mo-

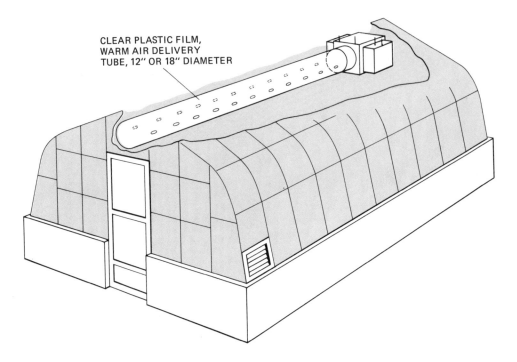

CLEAR PLASTIC FILM,
WARM AIR DELIVERY
TUBE, 12″ OR 18″ DIAMETER

This is the Janco jet-air heating and ventilating system from J. A. Nearing Co. It will heat and ventilate medium- and large-sized home greenhouses.

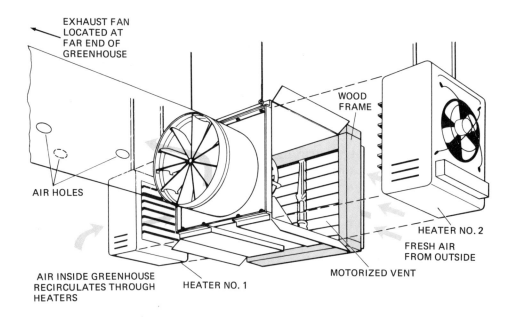

EXHAUST FAN
LOCATED AT
FAR END OF
GREENHOUSE

WOOD
FRAME

AIR HOLES

HEATER NO. 2

FRESH AIR
FROM OUTSIDE

MOTORIZED VENT

AIR INSIDE GREENHOUSE
RECIRCULATES THROUGH
HEATERS

HEATER NO. 1

torized outside-air inlet shutter and starts the exhaust fan. Air is then distributed through the polyethylene tube.

If you have a large glass greenhouse, in the 150- to 400-square-foot size, you can use a scaled-down fan-jet ventilating system. Here you can install a fan jet unit under a side bench, a few inches away from the end wall. A motorized shutter 15 inches square should be installed in the end of the greenhouse. Then a 12-inch circular polyethylene tube should be attached to the fan jet collar, and extended the full length of the greenhouse under the bench. The tube has exhaust holes punched at intervals, depending on volume of air to be circulated. The motor in the fan jet is 2-speed to provide air at either 780 or 1180 cubic feet per minute (cfm). Operating speed depends on the size of the greenhouse.

SHADING

In the winter, it seems, you can't get enough sun into the greenhouse. But in the summer, you should shade your plants to protect them from sunburn and to maintain optimum temperature and humidity condi-

Aluminum shades are adjusted from outside the greenhouse. Shades are available from greenhouse manufacturers in sizes for specific greenhouse models. (*Photo courtesy of Lord & Burnham*)

ALUMINUM ROLL SHADES

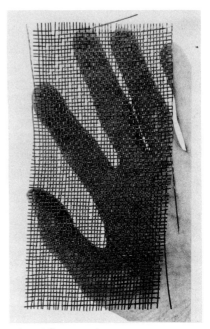

These plastic roll-up shades (left) are made of specially woven green-colored Saran and reduce light about 70 percent. A sample of 70 percent Saran is shown at right. But other weaves are available. Saran shades are less expensive than aluminum and are available for specific models of greenhouses and by the yard in several widths. (*Lord & Burnham*)

tions. There are many approaches to shading.

Commercial greenhouse operators, not too concerned about the external appearance of their plant factories, use shading paint that can be removed each fall. Shading paint comes in liquid and solid form, and in white or green. The density of shading is controlled by how much you dilute the paint before application. Most shading paints are diluted with water, but some long-lasting paints are diluted with benzene or gasoline. Before applying any shading paint, check to see how it can be removed in the fall.

Better shade control can be achieved with roll-up shades hung on the outside of the greenhouse. These shades are available with wood or aluminum slats, or they are made of woven Saran. One weave pattern gives the appearance of $\frac{5}{8}$-inch wide slats spaced $\frac{3}{8}$-inch apart.

Saran is also woven into shade cloth, which has a uniform weave, rather than a slatted one. This shade cloth comes in shade densities from 6 percent to 92 percent. It is hung inside the greenhouse.

Durable 8-mil vinyl plastic is also used for shading. This green film is cut to size and squeegeed to the inside of glass lights. Water surface tension holds it in place but allows it to be removed for storage in the fall and reused over several years.

COOLING

Natural ventilation, and even powered exhaust ventilation, may prove inadequate for summertime greenhouse cooling, even with shading. Evaporative coolers are made in many sizes. The cooler, enclosed in a steel cabinet located outside the greenhouse, consists of a fan blower that draws warm outside air through wet aspen wood pads; this causes the water to evaporate. The cool moist air produced is ducted into the greenhouse. This not only lowers the air temperature, but reduces watering requirements. Evaporative cooling can reduce the summer tem-

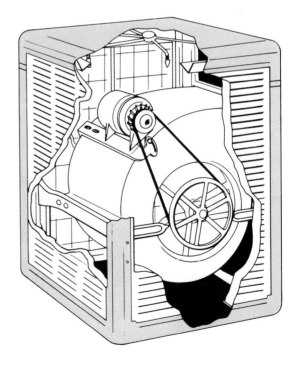

The Arctic Circle evaporative cooler contains fans that draw hot dry outside air through a thick layer of moistened aspen wood. This cools the air before it enters the greenhouse. Water is recirculated. Similar coolers are available from many suppliers.

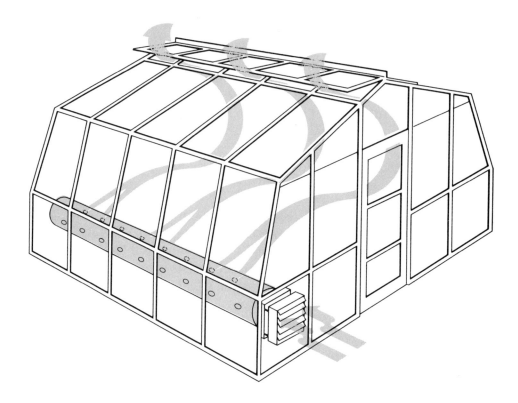

This Acme Fan Jet ventilating system is designed to cool only. It's suited for medium and large home greenhouses. Elements include roof vents, an inlet fan, and perforated plastic tubing located under one of the benches. (*Courtesy Lord & Burnham*)

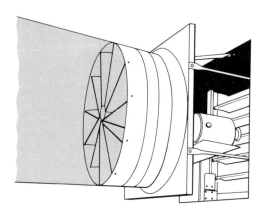

perature in a greenhouse by 20 to 30° F. For best operation, two-speed evaporative coolers require two thermostats set 4° F apart. When the greenhouse temperature rises to the lower setting, the cooler is turned on low-speed. If this rate of cooling proves inadequate and the temperature rises to the higher setting, the cooler is switched to high-speed operation.

THERMOSTATS

Thermostats sense air temperatures. When used to control electric heaters, their electric contacts must be capable of handling high-load currents. In addition, the thermostats must be equipped with sun shades to prevent false readings.

HUMIDITY

Most plants fare well with humidity between 30 and 50 percent. When artificial heat is being used in the winter, the humidity tends to drop far below this range. You can increase the humidity inside a greenhouse by any of a variety of devices that discharge water vapor into the air. You can increase the humidity simply by wetting down the walk and

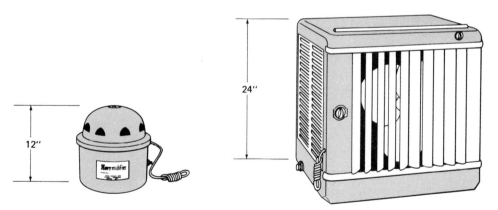

For small greenhouses, a "sick-room" humidifier, like the Herrmidifier, shown at left, may be sufficient to maintain optimal 30 to 50 percent humidity levels. But for a medium-sized or large home greenhouse, larger console models are needed. In selecting a humidifier, match its capacity to the size of your greenhouse. (*Courtesy J. A. Nearing, Inc.*)

floor. To mist a small greenhouse, you can use any kind of a home humidifier or a cool-vapor "sick room" humidifier. You can measure humidity level by means of a wet-bulb hygrometer (Taylor Instruments, $15). Or humidifiers can be controlled with a humidistat that senses the humidity of the air and turns the humidifier on and off. There are humidifiers made expressly for greenhouses, such as the Model 500 Herrmidifier that can automatically humidify up to 2,000 cubic feet of air to the level desired. The AC-powered Herrmidifier has an output of one quart of water per hour and costs about $135 with humidistat.

CARBON DIOXIDE

For photosynthesis and growth, plants require both light and carbon dioxide (CO_2). The level of CO_2 in normal outdoor air is 300 parts per million (ppm). On cold winter days, there is plenty of light, but with the greenhouse shut tight to conserve heat, the level of CO_2 in the greenhouse air can become deficient. Commercial greenhousemen have found that increasing the CO_2 level inside a greenhouse up to 3000 ppm justifies the expense in both plant yield and quality. You can enrich the CO_2 in your greenhouse with dry ice (frozen CO_2), compressed CO_2 in cylinders, or by burning alcohol (either methanol or ethanol) in a simple kerosene lantern. For a level of 2000 ppm in a 10 × 12-foot greenhouse, burn three ounces of ethanol or four ounces of methanol per eight-hour day when the greenhouse is lighted. A 2000 ppm CO_2 level in the greenhouse produces no hazard to anyone working inside.

6

How to Determine Your Heating Needs

Before you attempt to select an appropriate heater or heating system for your greenhouse, you should first determine how much heat the greenhouse will be losing. This is normally expressed in British thermal units per hour (Btuh). Note: One Btu equals the amount of heat necessary to raise one pound of water by one degree F.

A few greenhouse kit manufacturers publish heat-loss information that is valid for their greenhouses under specific conditions. But if you don't mind doing a little math, you can calculate greenhouse heat loss yourself no matter what the design or materials. And by doing the calculations yourself, you'll have a more thorough understanding of the overall heating task.

Kit manufacturers use any of several different methods of calculating heat loss. The method I'll describe is based on heat-loss factors for materials in the shells of homes. It is about the most reliable method. First, I'll explain the calculations in detail. Then, I'll show you some shortcuts and guide you through some practice exercises.

BASICS FOR HEAT FLOW CALCULATIONS

First, a couple of definitions: *Transmission loss* is the loss through the solid surfaces of all parts of the greenhouse, such as through glass, walls, floor, and through doors and vents. *Infiltration loss* is equal to

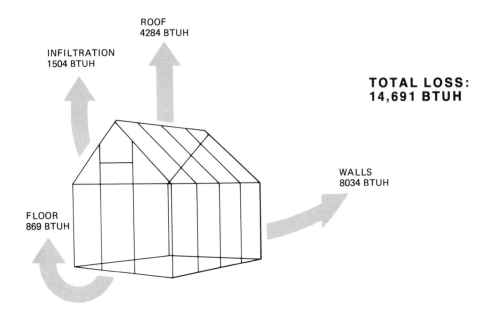

ROOF
4284 BTUH

INFILTRATION
1504 BTUH

**TOTAL LOSS:
14,691 BTUH**

WALLS
8034 BTUH

FLOOR
869 BTUH

**TOTAL LOSS:
11,527 BTUH**

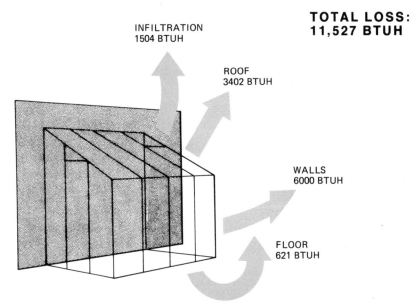

INFILTRATION
1504 BTUH

ROOF
3402 BTUH

WALLS
6000 BTUH

FLOOR
621 BTUH

These drawings show winter heat losses from even-span and lean-to greenhouses with the same floor space. Figures are based on a normal winter temperature of 10° F and a greenhouse temperature of 55° F. Conditions would be typical of those in the Philadelphia area. Heat could be supplied for the even-span by three 1500-watt electric heaters. The lean-to would need only two 1500-watt heaters.

the amount of heat needed to warm up the cold air leaking in through vent and door cracks. These losses must be calculated separately. In a mathematical formula, the equation looks like this:

$$\text{Total Btuh loss} = \text{transmission losses} + \text{infiltration losses}$$

Transmission Loss

The amount of heat that flows through a material, from one surface to the other, depends on that material's thermal conductivity. This is its k value, and it is expressed in Btuh per square foot, per degree F difference in temperature of the two sides, per inch of thickness.

Since most materials aren't used in one-inch thicknesses, another term, conductance, is often used. Conductance (C) is the same as k except it is given for common thicknesses of material, such as $\frac{1}{4}$-inch plywood or $\frac{1}{8}$-inch glass.

The glass of the greenhouse can be a lot colder than the air inside or outside. This difference in temperature between the air and the glass surface is surface conductance, usually called film conductance (f). The temperature difference between the outside face and the outside air is expressed as f_o (outside wall). And the difference between the inside surface and inside air is expressed as f_i (inside wall). These values may vary depending on whether the heat flow is up, down, or sideways, and depending on how hard the wind is blowing.

Enclosed, "dead" air has no air currents circulating in it. This is the air trapped between thermal panes and well-sealed storm windows. It has a thermal conductivity (a) that is lower than that of free-moving air. And for insulation purposes, the lower the thermal conductivity, the better. Thermal conductivity is the quantity of heat transferred (in Btuh) through one square foot of area per degree F of the surfaces bounding the air space, per hour.

Values for k, C, a, f_i, f_o for all common building materials, and many combinations of building materials, are listed in the *ASHRAE Handbook and Product Directory*, ASHRAE Inc., Publications Sales Dept., 345 East 47th St., New York, NY 10017. Knowing these values, you can calculate the amount of inside-air to outside-air transfer of any part of the greenhouse. The result is the coefficient of transmission, or the U factor for all of the parts.

$$U = \cfrac{1}{\cfrac{1}{f_i} + \cfrac{x}{k} + \cfrac{1}{C} + \cfrac{1}{a} + \cfrac{1}{f_0}}$$

In the above equation there is one term in the denominator for each element of the construction, with x/k or 1/C used as convenient for each layer.

Because of the number and letter mixture, this math sometimes leads to confusion unless R values (resistance to heat flow) are substituted:

$$R = \frac{1}{f_i}, R = \frac{1}{k}, R = \frac{1}{C}, R = \frac{1}{f_o}, \text{etc.}$$

Now, we can express U this way:

$$U = \frac{1}{R_1 + R_2 + R_3 + R_4 + R_5 + R_6} = \frac{1}{R_{T(Total)}}$$

Simply add up the Rs: The reciprocal (or the number 1 divided by R) is the U value. The table on page 144 provides U factors and R values of common greenhouse materials.

Now what's the transmission heat loss (thermal conduction) through each part of your greenhouse?

For each part,

Thermal Conduction Heat Loss $= A \times U \times (T_o - T_i)$ Btuh

In this calculation,

Btuh = British thermal units per hour

A = area in square feet

$U = 1/R_T$

T_o = outside temperature

T_i = inside temperature to be maintained

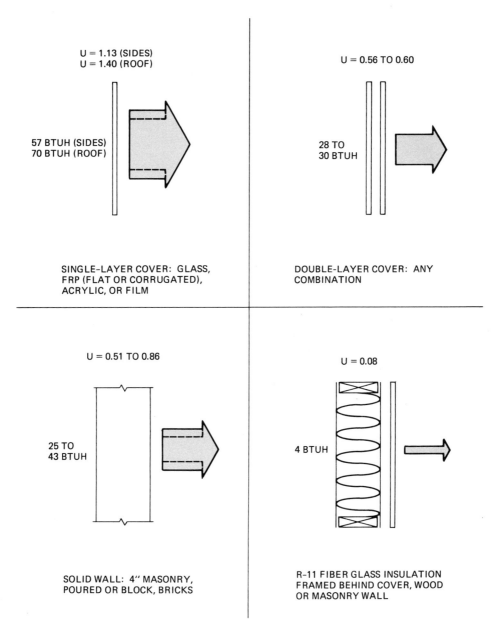

U = 1.13 (SIDES)
U = 1.40 (ROOF)

57 BTUH (SIDES)
70 BTUH (ROOF)

SINGLE-LAYER COVER: GLASS,
FRP (FLAT OR CORRUGATED),
ACRYLIC, OR FILM

U = 0.56 TO 0.60

28 TO
30 BTUH

DOUBLE-LAYER COVER: ANY
COMBINATION

U = 0.51 TO 0.86

25 TO
43 BTUH

SOLID WALL: 4″ MASONRY,
POURED OR BLOCK, BRICKS

U = 0.08

4 BTUH

R-11 FIBER GLASS INSULATION
FRAMED BEHIND COVER, WOOD
OR MASONRY WALL

These diagrams show winter heat transmission loss through various materials used in greenhouse construction. Infiltration loss is not included. In this case, the greenhouse temperature is assumed to be 55° F. Outside temperature is 5° F with a 15 mph wind. Btuh loss is noted per square foot of exposed area.

Let's assume that the greenhouse has tightly sealed glass all the way to the ground, that the frame is wood, that the floor is either dirt or gravel or a concrete slab at grade, and that the door and vents fit reasonably tightly.

Begin calculations by determining the area of the vertical and

U FACTORS AND R VALUES OF COMMON GREENHOUSE MATERIALS

Cover Materials	
Single glass, horizontal heat flow	U = 1.13
Single glass, vertical heat flow	U = 1.40
Double layer glass, 1″ to 4″ air spacing	U = 0.56
Corrugated fiberglass	U = 1.11
Flat rigid plastic	U = 0.92
Double layer flat rigid plastic	
$\frac{1}{4}$″ spacing	U = 0.53
$\frac{3}{4}$″ spacing	U = 0.48
1″ to 4″ spacing	U = 0.46
Single layer plastic film	U = 1.09
Double layer plastic film, pressurized	U = 0.70
Sand and gravel concrete, 8″ thick	R = 0.64
Concrete blocks, three oval cores	
4″ thick	R = 1.11
8″ thick	R = 1.72
12″ thick	R = 1.89
Common brick, per inch	R = 0.20
Face brick, per inch	R = 0.11
Stucco, per inch	R = 0.20
Asbestos-cement shingles	R = 0.03
Building paper	R = 0.06
Wood bevel siding, lapped	R = 0.81
Wood siding shingles, 16″, 7$\frac{1}{2}$″ exposure	R = 0.87
Insulation, glass fiber, mineral wool, etc.	
2″ to 2$\frac{1}{2}$″ thick	R = 7.00
3″ to 4″ thick	R = 11.00
5″ to 7″ thick	R = 19.00
Air spaces, heat flow horizontal, nonreflective	R = 7.00
system	R = 1.01
Air spaces, heat flow horizontal, reflective	
system	R = 3.48
Surface air films	
Inside (still air) heat flow up, nonreflective surface	R = 0.61
Heat flow horizontal, nonreflective surface	R = 0.68
Outside heat flow any direction, surface any position, winter	R = 0.17

near-vertical side and end walls. Next, determine the area of the greenhouse roof. Put both results in square feet. For quonset structures and domes, you will have to do a little guesswork.

The winter U value for vertical single glass (heat flowing horizontally) is 1.13. For horizontal, or nearly horizontal single glass (heat flowing upward), the winter U value is 1.4. (At the end of the calculations below, you'll find corrections to allow for other constructions.)

In our sample greenhouse shown in the accompanying drawing, surface area is computed this way:

Roof: two 6 × 5-ft. sections: $2(6 \times 5)$ $= 60$ sq. ft.
Walls:
   two 6 × 5-ft. ends $2(6 \times 5) = 60$
   two 8 × 5-ft. rectangular
     lower sections of
     front/back walls $2(8 \times 5) = 80$
                                     $= 164$ sq. ft.
   two 3 × 8-ft. triangular
     upper sections of
     front/back walls $2(\frac{1}{2} \times 3 \times 8) = 24$

This is the example greenhouse with dimensions used for calculations explained in the accompanying text.

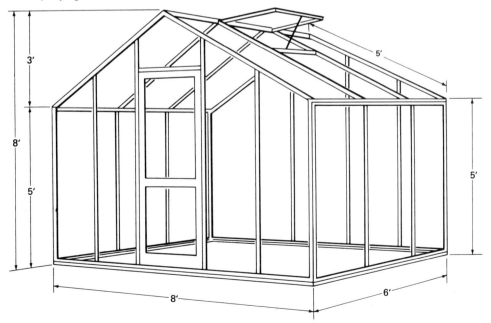

Assuming a 60°F inside temperature (T_i), and the coldest outdoors temperature (T_o) would be −5°F, the Btuh heat transmission losses (Q) are figured this way:

$$Q = \text{Area (A)} \times U \times (T_i - T_o)$$

For the walls, this becomes

$$164 \times 1.13 \times 65 = 12{,}046 \text{ Btuh}$$

For the roof it becomes

$$60 \times 1.4 \times 65 = 5{,}460 \text{ Btuh}$$

There is an additional heat loss through the floor of the greenhouse. Data for loss through dirt, gravel, or loose-laid brick floors is not available, but it isn't much different from the loss through an uninsulated concrete slab on grade. The heat loss through the portion of the floor within three feet of the edge varies with the inside-outside temperature differential. The heat loss from the center section is a constant 2 Btuh per square foot. (Continued on page 148.)

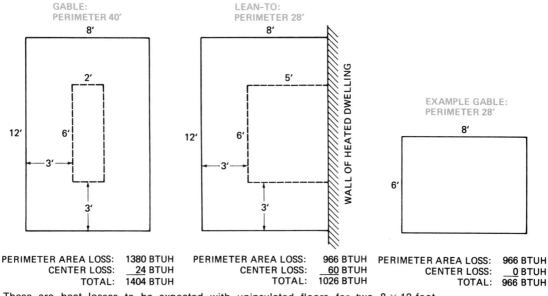

GABLE: PERIMETER 40'	LEAN-TO: PERIMETER 28'	EXAMPLE GABLE: PERIMETER 28'
PERIMETER AREA LOSS: 1380 BTUH	PERIMETER AREA LOSS: 966 BTUH	PERIMETER AREA LOSS: 966 BTUH
CENTER LOSS: 24 BTUH	CENTER LOSS: 60 BTUH	CENTER LOSS: 0 BTUH
TOTAL: 1404 BTUH	TOTAL: 1026 BTUH	TOTAL: 966 BTUH

These are heat losses to be expected with uninsulated floors for two 8 × 12-foot greenhouses and the gable greenhouse used as an example in the text. Here, it is assumed that the temperature difference inside and outside is 50° F. Computations are explained in the accompanying text.

COLDEST TEMPERATURE AND WINTER DEGREE DAYS

(Courtesy American Society of Heating, Refrigerating, and Airconditioning Engineers, Inc.)

Note: Heating degree days are used by the heating industry to indicate the number of degrees the daily average temperatures are below 65°F. A day with an average temperature of 45° has 20 heating degree days (65 − 45 = 20). If the average temperature for one day is 65°, that day has no heating degree days.

Location	Winter Coldest Temp.	Winter Degree Days	Location	Winter Coldest Temp.	Winter Degree Days	Location	Winter Coldest Temp.	Winter Degree Days
ALABAMA			Jacksonville	29	1200	**LOUISIANA**		
Birmingham	19	2600	Miami	42	200	Baton Rouge	25	1600
Decatur	15	3000	Orlando	33	800	Lake Charles	29	1400
Huntsville	16	3000	St. Petersburg	39	600	New Orleans	32	1400
Mobile	26	1600	Tallahassee	25	1400	Shreveport	22	2200
Montgomery	22	2200	Tampa	36	600	**MAINE**		
Tuscaloosa	19	2600				Bangor	− 8	8000
			GEORGIA			Lewiston	− 8	7800
ALASKA			Albany	26	1800	Portland	− 5	7600
Anchorage	−25	10800	Atlanta	18	3000			
			Columbus	23	2400	**MARYLAND**		
ARIZONA			Macon	23	2200	Baltimore	12	4600
Phoenix	31	1800	Savannah	24	1800	Hagerstown	6	5200
Tucson	29	1800						
			IDAHO			**MASSACHUSETTS**		
ARKANSAS			Boise	4	5800	Boston	6	5600
Fort Smith	15	3200	Coeur D'Alene	2	6600	Lawrence	− 3	6800
Little Rock	19	3200	Idaho Falls	−12	7200	Pittsfield	− 5	7600
Pine Bluff	20	2800	Pocatello	−8	7000	Springfield	− 3	6600
Texarkana	22	2600				Worcester	− 3	7000
			ILLINOIS					
CALIFORNIA			Aurora	−7	6600	**MICHIGAN**		
Bakersfield	31	2200	Chicago	−3	6600	Benton Harbor	− 1	6200
Fresno	28	2600	Elgin	−8	6600	Detroit	4	6200
Long Beach	36	1800	Peoria	−2	6000	Flint	− 1	7400
Los Angeles	41	2000	Rockford	−7	6800	Grand Rapids	2	6800
Oakland	35	2800				Holland	2	6400
Pasadena	36	2000	**INDIANA**			Saginaw	− 1	7000
Sacramento	30	2600	Bloomington	3	4800	Sault Ste. Marie	−12	9400
San Diego	42	1400	Columbus	3	5400			
San Francisco	35	3000	Fort Wayne	0	6200	**MINNESOTA**		
Stockton	30	2600	Muncie	− 2	5600	Duluth	−19	10000
			South Bend	− 2	6400	Minneapolis	−14	8400
COLORADO			Terre Haute	3	5400	Rochester	−17	8200
Boulder	4	5600				St. Paul	−14	8400
Denver	−2	6200	**IOWA**					
Fort Collins	−9	7000	Ames	−11	6800	**MISSISSIPPI**		
Leadville	−9	10600	Burlington	− 4	6200	Biloxi	30	1600
Pueblo	−5	5400	Cedar Rapids	− 8	6600	Jackson	21	2200
			Council Bluffs	− 7	6600	Natchez	22	1800
CONNECTICUT			Des Moines	− 7	6600	Vicksburg	23	2000
Bridgeport	4	5600	Iowa City	− 8	6400			
Hartford	1	6200	Sioux City	−10	7000	**MISSOURI**		
New Haven	5	5800				Hannibal	− 1	5400
Norwalk	0	5400	**KANSAS**			Kansas City	4	4800
			Dodge City	3	5000	St. Louis	4	5000
DELAWARE			Salina	3	5000			
Dover	13	4600	Topeka	3	5200	**MONTANA**		
Wilmington	12	5000	Wichita	5	4600	Billings	−10	7000
						Butte	−24	9800
DISTRICT OF COLUMBIA			**KENTUCKY**			Great Falls	−20	7800
Washington	16	4200	Covington	3	5200	Missoula	− 7	8200
			Lexington	6	4600			
FLORIDA			Louisville	8	4600	**NEBRASKA**		
Daytona Beach	32	800	Paducah	10	4000	Lincoln	− 4	5800
Fort Lauderdale	41	200				Norfolk	−11	7000

(Continued next page)

Location	Winter Coldest Temp.	Winter Degree Days	Location	Winter Coldest Temp.	Winter Degree Days	Location	Winter Coldest Temp.	Winter Degree Days
Omaha	-5	6600	**NORTH DAKOTA**			**TENNESSEE**		
Scottsbluff	-8	6600	Bismarck	-24	8800	Chattanooga	15	3200
			Fargo	-22	9200	Memphis	17	3200
NEVADA			Grand Forks	-26	9800			
Elko	-13	7400				**TEXAS**		
Las Vegas	23	2800	**OHIO**			Corpus Christi	32	1000
Reno	2	6400	Akron	1	6000	Dallas	19	2400
			Cleveland	2	6400	Houston	28	1400
NEW HAMPSHIRE			Lima	0	6000			
Concord	-11	7400	Sandusky	4	5800	**UTAH**		
Manchester	-5	7200	Toledo	1	5800	Ogden	7	8200
Portsmouth	-2	7200				Salt Lake City	5	6000
			OKLAHOMA					
NEW JERSEY			Oklahoma City	11	3200	**VERMONT**		
Atlantic City	14	4800	Stillwater	9	3800	Burlington	-12	8200
Newark	11	5000	Tulsa	12	3800	Rutland	-12	8000
Trenton	12	5000						
			OREGON			**VIRGINIA**		
NEW MEXICO			Baker	-3	7000	Lynchburg	15	4200
Albuquerque	14	4400	Eugene	22	4800	Norfolk	20	3400
Santa Fe	7	6200	Portland	21	4600	Richmond	14	3800
NEW YORK			**PENNSYLVANIA**			**WASHINGTON**		
Albany	-5	6800	Allentown	3	5800	Seattle	28	5200
Binghamton	-2	7200	Erie	7	6400	Spokane	-2	6600
Buffalo	3	7000	Harrisburg	9	5200	Walla Walla	12	4800
New York	12	5000	Philadelphia	11	4400			
Poughkeepsie	-1	6200	Pittsburgh	5	6000	**WEST VIRGINIA**		
Rochester	2	6800	Scranton	2	6200	Charleston	9	4400
Rome	-7	7400				Huntington	10	4400
Syracuse	-2	6800	**RHODE ISLAND**			Wheeling	5	5200
Watertown	-14	7200	Providence	6	6000			
						WISCONSIN		
NORTH CAROLINA			**SOUTH CAROLINA**			Eau Claire	-15	8000
Asheville	13	4000	Charleston	23	2000	Green Bay	-12	8000
Charlotte	18	3200	Greenville	19	3000	Madison	-9	7800
Greensboro	14	3800				Racine	-4	7400
Raleigh	16	3400	**SOUTH DAKOTA**					
Winston-Salem	14	3600	Aberdeen	-22	8600	**WYOMING**		
			Rapid City	-9	7400	Casper	-11	7400
			Sioux Falls	-14	8400	Cheyenne	-6	7400

The floor heat loss (Q) is:

$$Q = F \times P \times (T_i - T_o) + 2A$$

where:

Q = the heat loss through the ground

F = heat loss factor

P = length of the perimeter of the floor exposed to outside temperatures

T_i = temperature inside the greenhouse

T_o = outside temperature

A = area of the floor not included in the three-foot-wide perimeter

In our example greenhouse, P equals 28 feet, A equals 0. The heat loss factor F is found in the accompanying table (Heat Loss Factors). Heat loss through the greenhouse floor is:

$$0.60 \times 28 \times 65 + 0 = 1,092 \text{ Btuh}$$

HEAT LOSS FACTORS

Temperature Difference	F
50	0.69
55	0.66
60	0.63
65	0.60
70	0.58
75	0.57
80	0.55

Infiltration Loss

The infiltration loss through the cracks around the door and vents, and past the edges of the glass, must be added to the transmission losses calculated to get the total heat loss for the greenhouse.

Infiltration Heat Loss $= q \times .075 \times .24 \times (T_o - T_i) \text{ L Btuh}$

> $q =$ cubic feet of air leakage per hour per foot of crack
> $.075 =$ density of air in pounds per cubic foot
> $.24 =$ specific heat of air in Btuh per pound per degree F difference in temperature
> $L =$ length of crack in feet

The appropriate value for q without actual laboratory measurement is at best an approximation. But the following figures are reasonably accurate.

● Average wood vent without weatherstripping in wood framing is

39 cu ft/hr.
- Average weatherstripping wood vent in wood framing is 24 cu ft/hr.
- Average non-weatherstripped aluminum vent in aluminum framing is 72 cu ft/hr.
- Average weatherstripped aluminum vent in aluminum framing is 32 cu ft/hr.
- Average well-fitted door without weatherstripping is 80 cu ft/hr.
- Average well-fitted weatherstripped door is 48 cu ft/hr.

For a residence, infiltration losses would be calculated for windows and doors on only two sides. (For purposes of calculation, we have to figure that as much warm air is leaking out on two sides as cold air is infiltrating on the other two sides.)

To simplify our calculations, we have made an assumption that air infiltrates around the door and vents, and leaks out around the edges of the glass lites or plastic cover.

In our example, the door is 2 feet wide and 6 feet high, and the vent is 2-feet square, with average lites.

$$Q_{door} = 80 \times .075 \times .24 \times 65 \times 16 = 1{,}498 \text{ Btuh}$$
$$Q_{vent} = 72 \times 0.075 \times 0.24 \times 65 \times 8 = 674 \text{ Btuh}$$

The total heating load our example greenhouse could handle is as follows:

Walls	12,046
Roof	5,460
Floor	1,092
Infiltration	2,172
Total	20,770 Btuh

Such a heating load could be met with four 1500-watt electric space heaters.

In the calculations above, we assumed a wind velocity of 15 mph or less. With high wind velocities, heat loss will be greater. To calculate the capacity of your heating system for higher wind velocities, multiply the total heating load by the following factors:

WIND VELOCITY FACTORS

Wind Velocity (mph)	Factor
0–15	1.00
20	1.04
25	1.08
30	1.12
35	1.16
40	1.20

MATERIAL AND CONSTRUCTION FACTORS AFFECTING HEAT LOSS

Frame	Cover	Construction	Factor
Wood	Glass*	Well-sealed	1.00
Wood	Glass*	Reasonably sealed	1.13
Wood	Glass*	Poorly sealed	1.25
Metal	Glass*	Well-sealed	1.08
Metal	Fiberglass	—	1.00
Wood	Fiberglass	—	0.95
Metal	Film	One layer	1.00
Metal	Film	Two layers, pressurized	0.70

*Use same factors for rigid plastic panels installed in place of glass lights.

Calculations of total heat load have been made on the basis of an all-glass greenhouse. To allow for the effects of other types of materials, multiply the total heating load by the factors shown in the table titled "Material and Construction Factors Affecting Heat Loss."

ENERGY CONSERVATION

The need for energy-saving in greenhouse heating is critical. There are three things you can do to conserve energy:

- Double glaze by applying a second cover layer (not necessarily the same material as the existing cover).
- Seal all leaks to make the greenhouse as airtight as possible and thereby stop infiltration loss.
- Insulate lower walls and other walls of the greenhouse. (You could also partition part of the greenhouse in winter and only heat and use part of the total space.)

DOUBLE-GLAZING

Glass, plastic film, and sheets of plastic in the thicknesses used as greenhouse covers are by themselves essentially useless as thermal barriers. Any thermal resistance of single thickness cover is due almost entirely to the films of inert air that cling to its surfaces. The purpose of double-glazing is to double the number of surfaces, and trap a layer of dead air which does have thermal resistance. Up to an inch, effectiveness of this dead air increases with thickness. Then from one to four inches it remains about constant, and above four inches effectiveness goes down.

There are various schemes for double-glazing greenhouses. For glass, or rigid plastic-covered greenhouses, sheets of rigid acrylic (Plexiglas) or fiberglass can be attached to the inside of the bars for the gable, side, and roof. Make sure the plastic is sealed airtight against the bars. Remember that a lot of the effectiveness of any double-glazing is in eliminating infiltration losses. Polyethylene film can also be taped to the inside of the bars. A disadvantage of using plastic film is that any infiltration around rigid glazing will cause the film to flap, creating air currents in the "dead" air which reduces its insulating value.

Theoretically, double-glazing a greenhouse covered with glass or rigid plastic should result in about 50 percent reduction in heat loss through the cover. But in practice, a 25 percent savings is about the best you can achieve. Another effective method of reducing heat loss through glass is to apply adhesive-backed Aircap plastic bubble packing to the inside of the glass.

Film-covered greenhouses can be double-glazed by applying a second layer of film directly over the cover and pressurizing (inflating) the space between with a small electric blower. For details see Chapter 3. A two-layer pressurized film cover is claimed to result in a 40 percent saving in heating fuel as compared to the cost of heating a single-layer film covered greenhouse. This claim is based on the operation of a large commercial-size greenhouse. It's doubtful that such impressive results can be obtained in a home greenhouse.

Heat loss through the ground can be reduced several ways. If you plan to pour a concrete-slab floor, place a vapor barrier under the slab. And, no matter what kind of floor you plan, you can line the inside of a poured-concrete or cement-block foundation wall with rigid insulation. But the savings in heating cost with such a concrete floor might not be worth the operating inconvenience the rest of the year. Plain dirt floors covered with gravel or crushed rock are generally considered best for greenhouses because they won't dehumidify the air and dry out the plants. With a concrete floor, plants require more watering

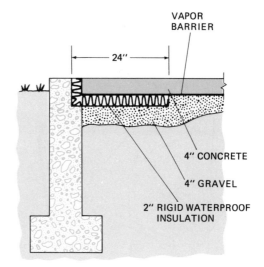

In these examples of greenhouse foundation insulation, construction **A** or **B** will cut floor heat loss to half that of an uninsulated floor. The slab should be sloped and a drain should be provided. Construction **C** is not as good because it has no vapor barrier.

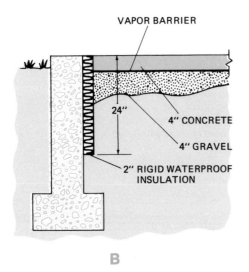

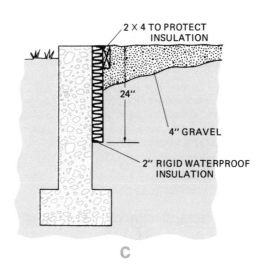

than with a dirt floor. And a concrete slab floor cannot provide the natural drainage of a dirt floor. The heat loss through a dirt floor runs only about 5 percent of the total greenhouse loss to begin with, yet every bit of savings counts.

Another way to reduce heat loss in the winter is to insulate the lower walls all around with rigid foam insulation or with fiberglass

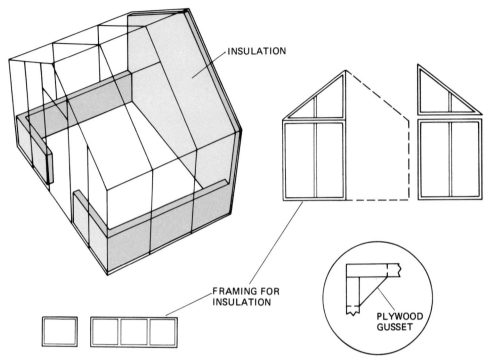

INSULATION

FRAMING FOR
INSULATION

PLYWOOD
GUSSET

Here are plans for frames that will support foil-backed, R-11 fiberglass insulation applied around the lower two feet of walls plus on all of the north-facing gable. The insulation panels can be removed in warm weather. You'll need two each of the frame sections shown. Use 1 × 4-inch pine and 24-inch-wide insulation. Make gusset corners with ¼-inch plywood to prevent tearing the foil.

batts held in place with tape and pastic film. Though some greenhouses are designed to be supported by low masonry walls below the glassed areas, a masonry wall of any kind is still a comparatively poor heat barrier. A two-foot masonry wall substituted for glass in our example greenhouse earlier would result in the following savings overall in Btuh load:

Original glass wall area =   164 sq. ft.

Masonry wall area = 2 sides (2 × 6) +
    back (2 × 8) +
    front less door (2 × 6) =   24 + 16 + 12 =   52 sq. ft.

Glass wall area (164 sq. ft.) less
   masonry wall area
   (52 sq. ft.) =   112 sq. ft.

New glass wall area heat loss =
   Area × U factor ×
   $(T_i - T_o) =$ $112 \times 1.13 \times 65 =$ 8,226 Btuh

The masonry wall area loss (U factor) is computed

$$U = \frac{1}{R} \times \frac{1}{f_i} \times \frac{1}{f_o} = \frac{1}{1.11 + 0.17 + 0.68} = 0.51$$

The loss through the concrete block wall is

Area $\times$ U factor $\times (T_i \times T_o) =$ $52 \times 0.51 \times 65 = 1,724$ Btuh

Whereas the loss through a comparable area of glass would be

$52 \times 1.13 \times 65 = 3,819$ Btuh

Foil-backed fiberglass batt insulation can be installed against the lower walls of a greenhouse. All cracks and openings should be plugged with loose insulation.

Thus, the block wall would result in savings of 2,095 Btuh, or an overall 10 percent savings. Unfortunately, to achieve this savings in heating cost, you must sacrifice a significant amount of growing area.

As an alternative, you could tape four inches of R-11 fiberglass insulation to the inside of the lower two feet of the glass area. (Also assume a 1×4 inch ledge around the top for neatness.) The calculations would be as follows:

$$U = \frac{1}{r + f_o + f_i}$$

$$= \frac{1}{11 + 0.17 + 0.68} \quad \text{(ignoring the R of glass)}$$

$$= \frac{1}{11.85}$$

$$= .084$$

$$\text{Heat Loss} = 52 \times 0.084 \times 65 = 285 \text{ Btuh}$$

The overall heat loss savings by this insulation amounts to 3,535 Btuh—a savings of 17 percent.

A better alternative would be to insulate the north wall in winter—in our example greenhouse, one gable end—using foil-covered insulation so that light would be reflected back in. That way you could save 17 percent, without sacrificing any growing area.

PART II

Garden Shelters & Sheds

This white-painted redwood gazebo has the traditional octagonal shape. Both the view of the gazebo and the view from the gazebo are attractive. And the gazebo is sized appropriately for its setting. (*Courtesy California Redwood Assn.*)

7

Gazebos for Shelter and Gracious Living

A gazebo is traditionally an open, summer structure designed for conversation, dining, reading, and for plain old relaxing. The design originated in Holland where gazebos were built primarily of stone. In the New World, wood became the most common construction material. Gazebos can be built on natural or man-made hillocks to exploit cool breezes and to elevate the occupants above low-flying insects that become a nuisance on summer evenings. This principle is evident today in the siting of gazebos at Colonial Williamsburg.

Today, the term gazebo may also apply to outdoor structures that are essentially enclosed. Whether a gazebo is enclosed or open, its sides are normally of painted wood, but weathered or stained wood, wrought ironwork, and even stone and masonry are used. Heavy timbers look clumsy in a gazebo. A gazebo should be light-structured.

When planning a gazebo, it's best to think small. A gazebo should always look detached, cozy, and smaller than its immediate surroundings—especially in relation to your home. It's not intended to serve as a porch. If it is spacious or sprawling it begins to look more like a summer house, which is something else.

The dividing line between a gazebo and a garden shelter is hazy. Decorative garden shelters may resemble gazebos in many ways, but such shelters are usually planned with shelter rather than gracious living as the prime consideration. The gazebo can be erected

This hexagonal gazebo measures only 6 feet across. A shingled roof could easily be substituted for the slatted roof shown. Construction is redwood, left unfinished. (*Courtesy California Redwood Assn.*)

as a permanent structure, or it can be put up in sections that can be disassembled when you decide to move.

The usual floor plan of a gazebo is octagonal, hexagonal, rectangular, or square. But there's no reason a gazebo couldn't be five-, seven-, or nine-sided, or round. A gazebo layout can be designed to fit on any terrain or location.

Seating can be either built-in (almost mandatory in small gazebos). Or you can simply set up lawn or patio furniture inside. Bulky redwood lawn furniture tends to look too heavy in a small gazebo.

BASIC GAZEBO CONSTRUCTION

Most gazebos are of post-and-beam construction. Posts are set into the ground, or anchored to concrete footings or a concrete slab. Redwood and pressure-treated wood posts can be set directly into the ground, and firmly tamped earth will provide enough rigidity. However, if the soil is sandy or unstable, tamp the earth around the bottom of the pole and then pour a concrete collar. You can rent a tamping bar, or you can simply use the end of a 2 × 4.

Post depth should take into account factors such as wind, total number of posts, soil conditions, and the local frost line. A gazebo 8 to 10 feet high requires posts 3 feet deep for most wind and soil conditions. With posts set in the ground, the floor can be wood decking, gravel, paving blocks, earth, grass, artificial turf, or indoor-outdoor carpeting.

POPULAR GAZEBO SHAPES

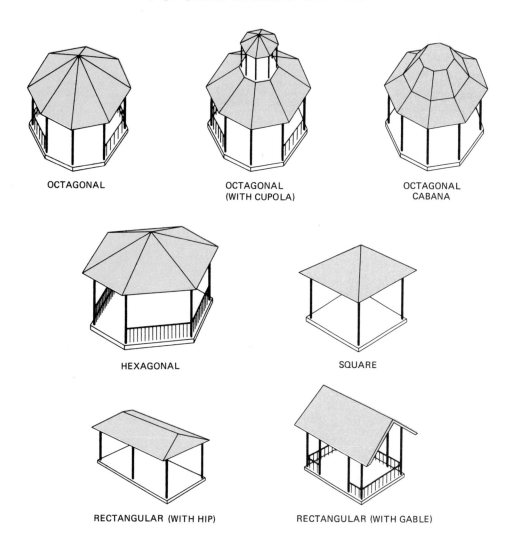

OCTAGONAL

OCTAGONAL
(WITH CUPOLA)

OCTAGONAL
CABANA

HEXAGONAL

SQUARE

RECTANGULAR (WITH HIP)

RECTANGULAR (WITH GABLE)

If you want a concrete-slab floor, the posts can be anchored to the slab rather than set in the ground. It is important that the posts be *positively* anchored to the concrete. There is a variety of post anchors available that are imbedded in the concrete when it is poured. The posts are then bolted to the anchors. Angle fasteners (mending angles) can also be used. But they take longer to install. They don't provide as secure

METHODS OF ANCHORING POSTS

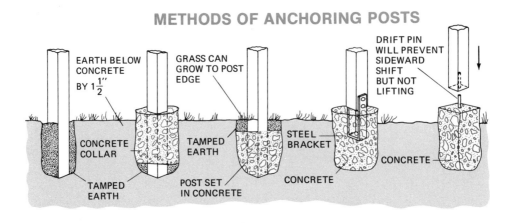

EARTH BELOW
CONCRETE
BY $1\frac{1}{2}''$

CONCRETE
COLLAR

TAMPED
EARTH

GRASS CAN
GROW TO POST
EDGE

TAMPED
EARTH

POST SET
IN CONCRETE

STEEL
BRACKET

CONCRETE

DRIFT PIN
WILL PREVENT
SIDEWARD
SHIFT
BUT NOT
LIFTING

CONCRETE

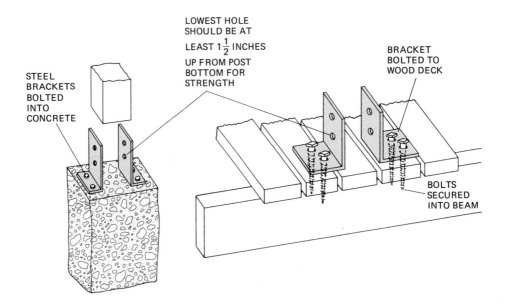

STEEL
BRACKETS
BOLTED
INTO
CONCRETE

LOWEST HOLE
SHOULD BE AT
LEAST $1\frac{1}{2}$ INCHES
UP FROM POST
BOTTOM FOR
STRENGTH

BRACKET
BOLTED TO
WOOD DECK

BOLTS
SECURED
INTO BEAM

an anchor. And they restrict the floor space more. But angle fasteners can be used in case you don't set anchors when concrete is poured. J-bolts, used to anchor house framing sills to the foundation, do not make satisfactory post anchors because of the near impossibility of attaching them to the end of a post. They can be used as drift pins if there is some other means of hold-down used. Another anchoring method

SQUARE WOOD POST   CONCRETE-BLOCK FOUNDATION   ROUND WOOD POST

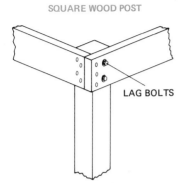

LAG BOLTS

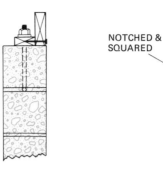

NOTCHED & SQUARED

Here are a few ways of attaching sill plates to posts. Note the notched and squared sides of the wood posts.

POST ANCHORS

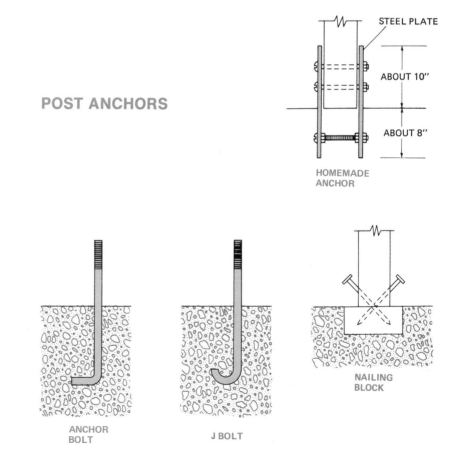

STEEL PLATE

ABOUT 10"

ABOUT 8"

HOMEMADE ANCHOR

ANCHOR BOLT

J BOLT

NAILING BLOCK

employs a wood block larger than the cross-section of the post set in the concrete, with the post toe-nailed to the block. This anchor is far less secure than a post bolted to a metal anchor that is imbedded in the concrete. You can make an anchor by using two steel plates spaced apart by a thrubolt so they will fit against the sides of the post. Imbed in concrete the ends of the plate containing the bolt. And bolt the post between the protruding plate ends.

Beams support the roof rafters and tie the tops of the posts together. It is important that the beams are securely attached to the post tops. Simple nailing as is done in house construction is not enough. (This nailing works in house construction because sheathing and diagonal bracing are added, and this locks the nailed pieces together, preventing racking.) Preferably, the beams should be directly on top of the posts but a gazebo design might make this impossible. The beams must be fastened firmly to the post, and they must also be fastened firmly to each other where they butt at a corner, because the

SIMPLE POST AND BEAM JOINTS

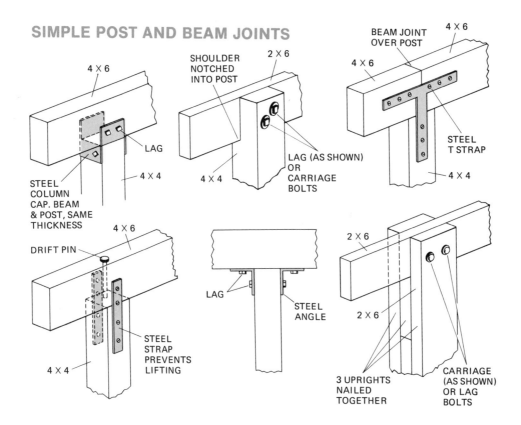

WOOD FLOORING

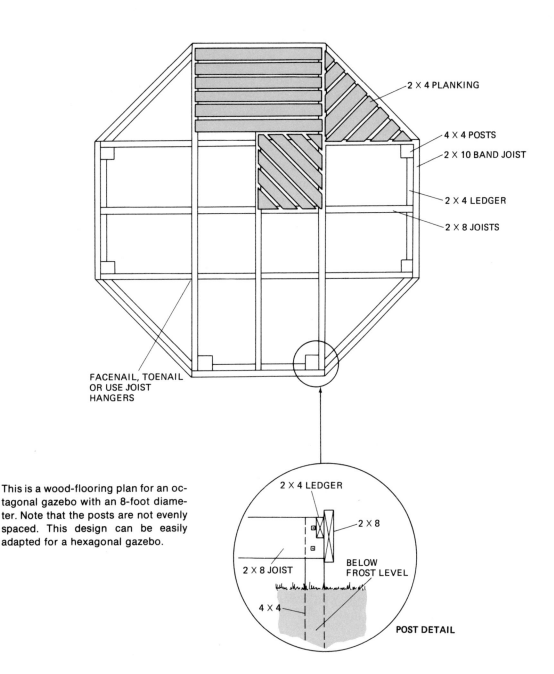

2 X 4 PLANKING

4 X 4 POSTS

2 X 10 BAND JOIST

2 X 4 LEDGER

2 X 8 JOISTS

FACENAIL, TOENAIL
OR USE JOIST
HANGERS

This is a wood-flooring plan for an octagonal gazebo with an 8-foot diameter. Note that the posts are not evenly
spaced. This design can be easily
adapted for a hexagonal gazebo.

2 X 4 LEDGER

2 X 8

2 X 8 JOIST

BELOW
FROST LEVEL

4 X 4

POST DETAIL

beam-to-beam joint is in tension. For added strength, you can mount metal plates over the joints.

Beams can be attached to the posts with metal post caps, or shouldered into partial lap joints. Although post caps are designed for use with beams the same thickness as the posts, they can be used in odd angle joints on gazebos with more than four sides and then fitted to 2-inch dimension lumber with filler blocks to make a secure connection. At intermediate post and beam joints, single or double wood cleats can be used. Wood trim pieces are often added to cover the metal fasteners.

Rafters support the roof. They can be strictly utilitarian and hidden under the finished roof, or they can be a visible and decorative part of the roof. Rafters should be notched where they connect with the beams for added strength. If no rafter overhang is desired, the rafters can be attached to the beams with rafter hangers, which are preferable to ledger strips that are only nailed to the beam to support the rafter. The inner ends of the rafters must be tied in with a steel plate because a mitered and nailed joint is not strong enough. The gazebo will be a lot cooler if a ventilating cupola is provided. For this, the upper ends of rafters are connected to a frame for the cupola.

If the gazebo is to provide shelter from inclement weather, the roof has to be solid. If it is only to provide shade, the roof can be an open latticed or slatted roof. Cedar shakes or shingles, or shingles of asphalt or fiberglass, make good gazebo roofs, as does a double layer of bamboo or reed fencing. The finished roofing and its supporting lath or plywood can be laid over the rafters for a more or less conventional roof, or they can be supported on ledgers nailed to the sides of the rafters so that the rafters are exposed above the roof.

Roofing nails protruding through plywood may be a normal sight in your attic, but they lend a barn-like quality to a gazebo. For a finished look, a false ceiling should be installed. This is most easily accomplished before you do the roof.

Gazebos are basically open structures, with the spaces between the posts filled at most with railings, trellis work, or gingerbread. For privacy, protection from prevailing wind, or screening an unattractive view, one or more sides can be solid or can be covered with a roll-up vinyl or bamboo shade. Part of a gazebo could also be made solid for storage or poolside dressing rooms. The possibilities are endless.

OCTAGONAL GAZEBO POST-AND-BEAM JOINTS

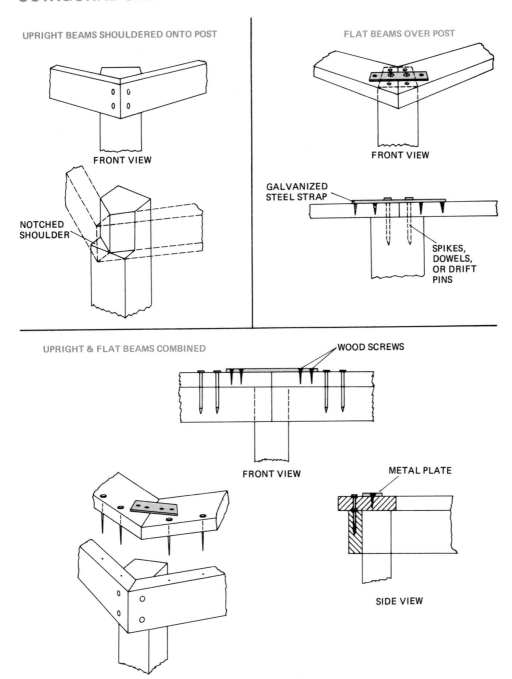

UPRIGHT BEAMS SHOULDERED ONTO POST

FRONT VIEW

NOTCHED
SHOULDER

FLAT BEAMS OVER POST

FRONT VIEW

GALVANIZED
STEEL STRAP

SPIKES,
DOWELS,
OR DRIFT
PINS

UPRIGHT & FLAT BEAMS COMBINED

WOOD SCREWS

FRONT VIEW

METAL PLATE

SIDE VIEW

SMALL HEXAGONAL GAZEBO

Open slat roof is shown. Roof could be sheeted with plywood and covered with asphalt or fiberglass shingles. Or the roof could be lathed and covered with cedar shingles.

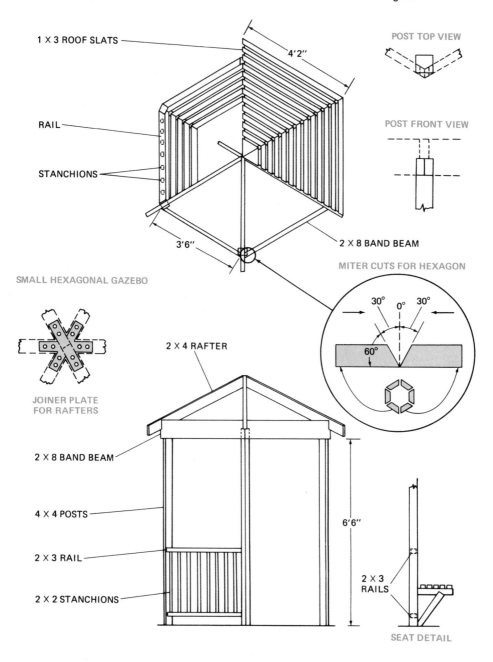

1 × 3 ROOF SLATS

4'2"

POST TOP VIEW

RAIL

POST FRONT VIEW

STANCHIONS

3'6"

2 × 8 BAND BEAM

SMALL HEXAGONAL GAZEBO

MITER CUTS FOR HEXAGON

30°   0°   30°

60°

JOINER PLATE FOR RAFTERS

2 × 4 RAFTER

2 × 8 BAND BEAM

4 × 4 POSTS

6'6"

2 × 3 RAIL

2 × 3 RAILS

2 × 2 STANCHIONS

SEAT DETAIL

OCTAGONAL GAZEBO WITH CUPOLA

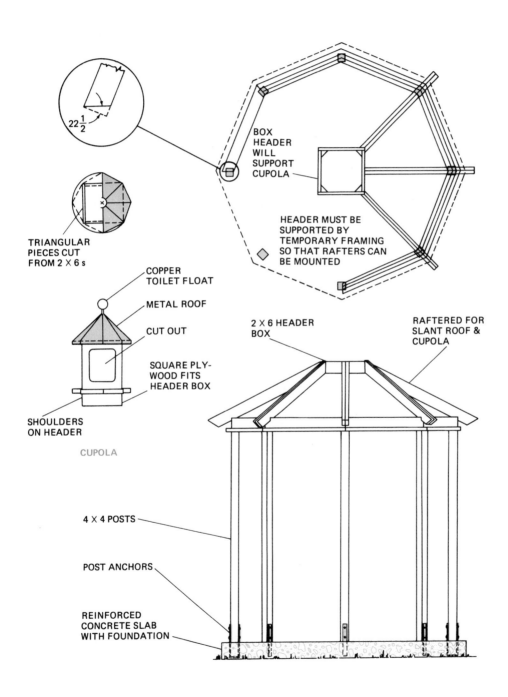

$22\frac{1}{2}$

BOX
HEADER
WILL
SUPPORT
CUPOLA

HEADER MUST BE
SUPPORTED BY
TEMPORARY FRAMING
SO THAT RAFTERS CAN
BE MOUNTED

TRIANGULAR
PIECES CUT
FROM 2 × 6 s

COPPER
TOILET FLOAT

METAL ROOF

CUT OUT

SQUARE PLY-
WOOD FITS
HEADER BOX

SHOULDERS
ON HEADER

CUPOLA

2 × 6 HEADER
BOX

RAFTERED FOR
SLANT ROOF &
CUPOLA

4 × 4 POSTS

POST ANCHORS

REINFORCED
CONCRETE SLAB
WITH FOUNDATION

OCTAGONAL GAZEBO WITH SOLID ROOF

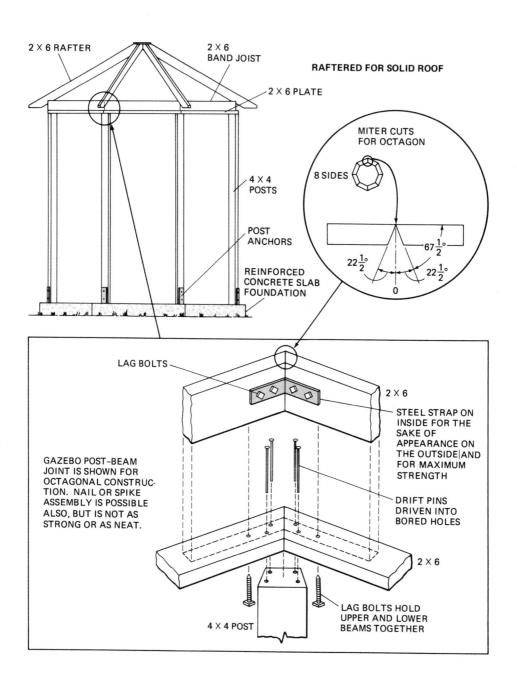

2 × 6 RAFTER

2 × 6 BAND JOIST

RAFTERED FOR SOLID ROOF

2 × 6 PLATE

MITER CUTS FOR OCTAGON

8 SIDES

$67\frac{1}{2}°$

$22\frac{1}{2}°$

$22\frac{1}{2}°$

0

4 × 4 POSTS

POST ANCHORS

REINFORCED CONCRETE SLAB FOUNDATION

LAG BOLTS

2 × 6

STEEL STRAP ON INSIDE FOR THE SAKE OF APPEARANCE ON THE OUTSIDE AND FOR MAXIMUM STRENGTH

GAZEBO POST–BEAM JOINT IS SHOWN FOR OCTAGONAL CONSTRUC- TION. NAIL OR SPIKE ASSEMBLY IS POSSIBLE ALSO, BUT IS NOT AS STRONG OR AS NEAT.

DRIFT PINS DRIVEN INTO BORED HOLES

2 × 6

LAG BOLTS HOLD UPPER AND LOWER BEAMS TOGETHER

4 × 4 POST

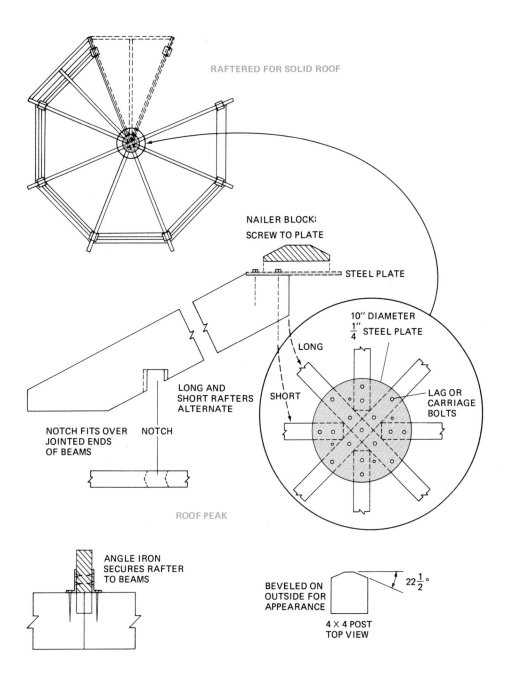

RAFTERED FOR SOLID ROOF

NAILER BLOCK:
SCREW TO PLATE

STEEL PLATE

10" DIAMETER
$\frac{1}{4}$" STEEL PLATE

LONG

LONG AND
SHORT RAFTERS
ALTERNATE

SHORT

LAG OR
CARRIAGE
BOLTS

NOTCH FITS OVER
JOINTED ENDS
OF BEAMS

NOTCH

ROOF PEAK

ANGLE IRON
SECURES RAFTER
TO BEAMS

BEVELED ON
OUTSIDE FOR
APPEARANCE

$22\frac{1}{2}$°

4 X 4 POST
TOP VIEW

TINY, SQUARE GAZEBO

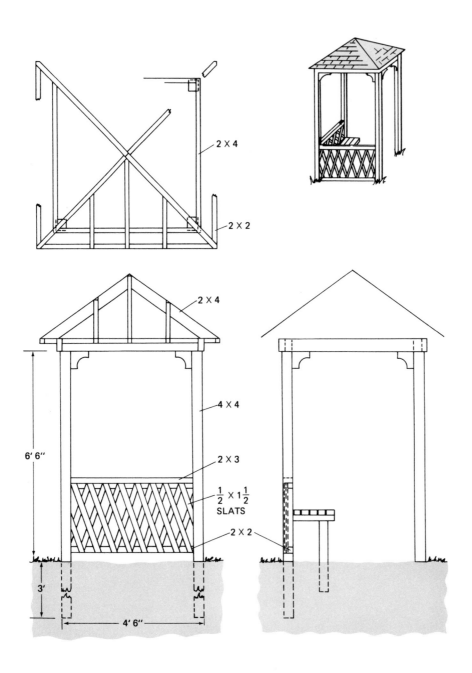

2 X 4

2 X 2

2 X 4

4 X 4

2 X 3

$\frac{1}{2}$ X $1\frac{1}{2}$
SLATS

2 X 2

6' 6"

3'

4' 6"

CURVED-ROOF GAZEBO

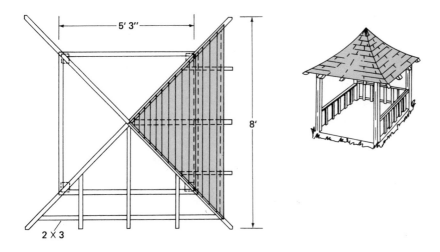

The curved roof of this gazebo can be covered with flexible shingles or with slats.

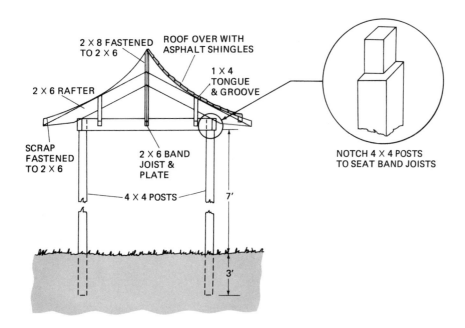

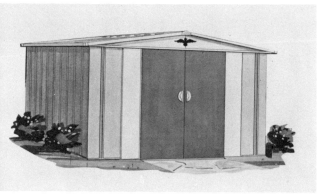

174

8

Storage Sheds for Every Purpose

There comes a time in every homeowner's life when the gradual accumulation of outdoor tools and equipment no longer fits under the car port or in garage, with the car there too. One solution is a yard storage shed.

Once you start considering a storage shed, you begin to realize it can be used for things in addition to storage. A storage shed can be used simply as a substitute attic or basement for locked-up year-round storage. Or it can be used seasonally. In summer it can shelter lawnmower, barbeque, garden tools, and patio chairs. In winter it can house sleds, snowblower, and snowmobile. Just about all kinds of sheds can be used for garden work centers. Many can be used as small workshops, poolside dressing rooms, or even as guest houses.

Before trying to decide how much storage shed you need and optional uses for it, make up a list of everything that you might want to store in a shed that is now stashed away in your attic, cellar, crawl space, garage, carport, closets, spare room, as well as lying around the yard. Will keeping everything in one place create a bigger problem than you now have with it scattered? How you will use a storage shed should be a big factor in determining its size, location, and design. Maybe several storage sheds would be better than just one. Each could be located conveniently near its point of primary use—garden,

A yard storage shed can be used for many things. Many people find that a shed allows them to clean out an over-stuffed garage well enough to fit the car back in the garage again. (*Courtesy JER Manufacturing Co.*)

This little red barn is painted white. Plans and kits for it are available from many sources. The barn shown comes in a kit of pre-cut framing parts sold through lumber yards. Siding and roofing are purchased separately. Lumber yards often advertise the kit and additional materials as a package. (*Courtesy JER Manufacturing Co.*)

Thin steel storage shed has sliding doors and ridge-formed side and roof panels for strength. (*Courtesy Robco.*)

COMMON SHED STYLES

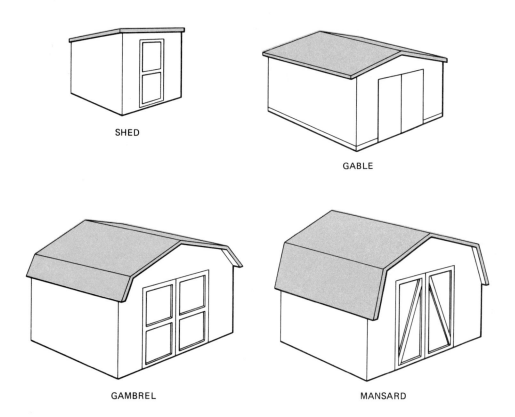

SHED

GABLE

GAMBREL

MANSARD

woodpile, pool. Gardeners can use a shed for potting or for storage of wheelbarrow, tools, flats, mulches, potting soil, and fertilizer.

Naturally, a shed can have multiple uses. A partition can divide a shed into a storage and a work area, or into poolside storage and dressing rooms, or into two dressing rooms.

There are basically five approaches to do-it-yourself storage sheds. You can buy a kit for a steel shed. Or you can buy a prefabricated wood shed. Or you can buy a kit for a wood shed that comes with the frame pieces cut to size but without siding and roofing. Or you can buy plans and build a shed yourself. Or you can design and build your own.

Steel storage sheds are the least expensive, and the easiest of the kits to erect. Assembly is with wrench and screwdriver. Solid anchor-

Here's a rustic shed in pine that comes in major sub-assemblies. It can be erected in a day. (*Courtesy Walpole Woodworkers Co.*)

Plans for this mini storage barn are available from the American Plywood Association.

ing is necessary for all metal sheds because they are lightweight. Wind can damage them easily unless they are firmly attached to the ground.

The best base for a metal storage shed is a concrete slab, but putting in a slab can lead to zoning difficulties. On a slab, the building becomes permanent. But without the slab, it may be considered temporary and you often will have more leeway in putting it up. We covered many of these factors in Chapter 1. A variety of anchors is available. If you are going to store tools in the shed, you should have some kind of flooring—concrete slab, wood, chipped board, or steel. Some

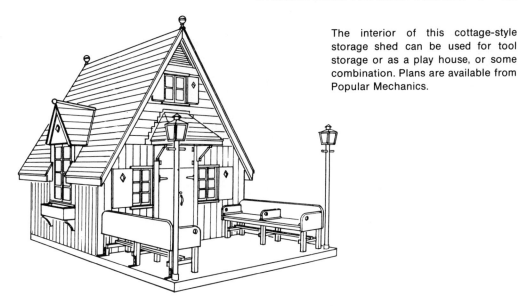

The interior of this cottage-style storage shed can be used for tool storage or as a play house, or some combination. Plans are available from Popular Mechanics.

kits provide the flooring. Others don't. If the floor is wood, it should be elevated off the ground a few inches and ventilated underneath.

Assembly of a typical steel shed kit begins with lag-bolting perimeter floor frames to a concrete base with masonry anchors and then screwing the frame to a wood base (see Chapter 3), or screwing the frames to the tops of driven stakes. At this point it is important that the frame is square and level, or you will have difficulty with the rest of the assembly. The walls and corners consist of interlocking corrugated steel panels that are one to two feet wide. These are attached to the floor frame with self-tapping screws. Gables and side beams are attached next. Then comes the sliding track for the door, Next come the ridge beam and roof braces. Corrugated roof panels 24-inches wide are then bolted on. The shed is completed by installing trim and hanging the doors on the track.

OTHER SHEDS

Other styles of wooden yard storage sheds are available besides the little red barns. Styles include plain rustic buildings with board and batten siding, carriage houses, and cottages complete with Victorian gingerbread.

A prefab shed with an integral floor does not need a slab foundation. It can be mounted on piers. For an 8 × 10-foot shed, piers are

In the exploded view of this steel storage shed, it's apparent why no cutting or fitting is required. All you need are a screwdriver and wrenches. (*Courtesy Roper Eastern*)

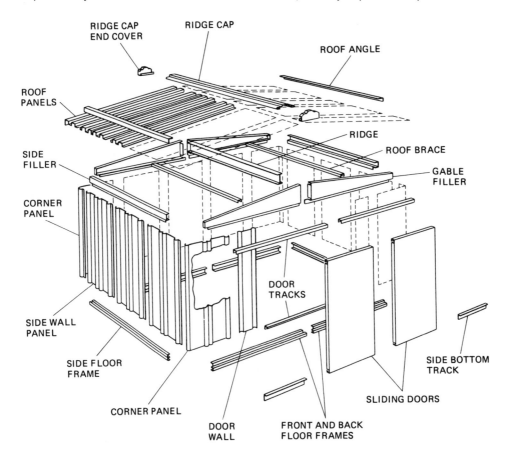

Plans for these storage sheds are available from the American Plywood Association. The shed at left has a 6 × 8-foot floor area. The model at right has an 8 × 12-foot floor.

needed at the corners, and at the center of the long sides. For reasonable stability on a level grade, a minimum of two blocks should be used, with 3 inches of the top block exposed above grade. More blocks can be stacked for sloping grades. It is important that the top surfaces of all the piers be level, and you may need shims for this. You can achieve better stability if the piers extend below the frost line; in this case, it is easier to pour concrete piers. But if you go with blocks, only the exposed blocks need to be mortared together.

In a typical shed, the floor comes in two halves which are placed on the foundation blocks and spiked together. They should be shimmed between the floor framing and the blocks if necessary until absolutely level. The back wall is then stood on the floor and braced temporarily. Next step is to raise a side wall and bolt the walls together. When all walls are bolted, the roof sections are raised and bolted to the walls. Then tie beams are nailed between ends of front and rear rafters to prevent spreading. The only extended amount of work in erecting the prefab shed is shingling the roof. Last, bottom ends of the wall boards are nailed to the floor framing, and the trim is nailed on.

This big red barn can hold anything from horses to a good-sized boat with room to spare. (*Courtesy JER Manufacturing Co.*)

Even this 5 × 6-foot metal storage shed can help you remove a lot of items from the garage. (*Photo is courtesy of Robco.*)

OTHER STORAGE

Walk-in sheds are not the only practical means of storing things. Not every yard has room for a large single shed, and you just might rather use the available space for something else anyway. The solution is specialized storage structures which can be placed about the yard where convenient. One advantage of spreading your storage in several places is that you will spend less time moving things around in order to reach tools or equipment stored up against back walls. Of course you will have to remember where you keep things.

ARCADIA SHEDS PLANS

Arcadia plans show you how to build any of three saltbox sheds as well as woodshed attachments. First you pre-fabricate individual panels and then bolt them together. The sheds sit on concrete piers. Framing consists of 2 × 6s and 2 × 4s. Siding is 1-inch board-and-batten pine. Maxi Shed 812 measures 8 × 12 feet and has an attached 5 × 6-foot woodshed. The mini sheds measuring 6 × 8 feet come in two versions: Mini 68, with closed front and door, and Mini 68WS, with open woodshed front. Both minis feature 5 × 6 woodshed wings. Plans are scaled ½-inch to the foot. Plans for the Maxi are $7 and for each mini, $6. (Address in Appendix.)

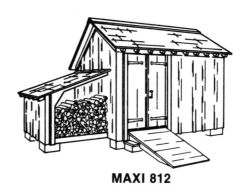

MAXI 812

MINI 68

MINI 68SW

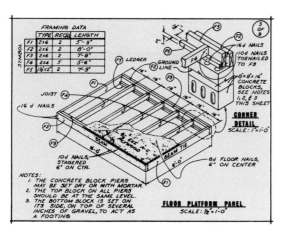

This is a sample of one of nine plan sheets greatly reduced in size.

CLUSTER SHED'S VERSATILE SHELTER

This versatile 12 × 16-foot shelter can serve as a gazebo, woodshed, workshop, garden storage center, or as one of several components of a large cluster-arranged home. Here 6 × 6 posts and 4 × 6 rafters of white pine are connected by means of mortise and tenon joints held fast by oak pegs. Assembly is best done in the barn-raising tradition with plenty of strong helpers.

CLUSTER MORTISE These details show the mortise and tenon joints used for a post (left) and for a rafter peak. Steel spikes are used at the rafter plate, however. In olden days these joints were fashioned with saws, drills, axes, and chisels. Today, precut versions are machined for precise, tight fits.

CLUSTER ROOF DETAIL Note the overlap of split cedar shingles and roofing felt laid over spaced roof boards. The first course is double.

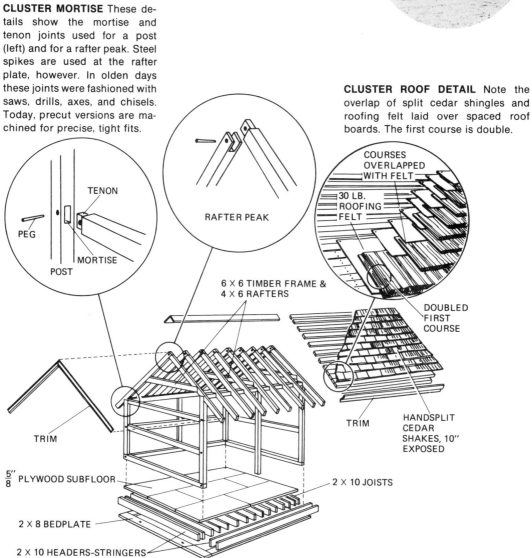

PEG

TENON

MORTISE

POST

RAFTER PEAK

COURSES OVERLAPPED WITH FELT

30 LB. ROOFING FELT

DOUBLED FIRST COURSE

6 × 6 TIMBER FRAME & 4 × 6 RAFTERS

TRIM

HANDSPLIT CEDAR SHAKES, 10" EXPOSED

TRIM

$\frac{5}{8}$" PLYWOOD SUBFLOOR

2 × 10 JOISTS

2 × 8 BEDPLATE

2 × 10 HEADERS-STRINGERS

GARDEN SHED & GAZEBO

This structure was designed and built by how-to writer John Capotosto in his own backyard. It features an octagonal gazebo front merged into a storage shed with entry on the backside. The shed portion measures about 8 × 10 feet, and the gazebo portion extends forward from the shed about 7 feet. People passing by Capotosto's house see the shed and inquire where they might buy plans. Usually, Capotosto has a set of plans on hand because he sells them through his J.C. Armor Company, which specializes in mail-order items for the home craftsman. Plans are $4.50 from J. C. Armor Co., Inc., Dept. PSB, Box 290, Deer Park, NY 11729. (*Photo courtesy Masonite Corp.*)

SIDING

Siding for storage sheds and other yard structures is classed as either structural or nonstructural, depending on whether the siding contributes much to the rigidity of the structure. Plywood siding, for example, can add enough rigidity to the frame so that diagonal bracing is not necessary. Aluminum siding, on the other hand, is only hung on the building and not nailed securely as wood siding is. The choice of siding material determines the framing and sheathing requirements of the yard structure.

Plywood siding, in thicknesses of $\frac{1}{4}$, $\frac{3}{8}$, $\frac{1}{2}$ or $\frac{5}{8}$ inch can be nailed directly to studs, eliminating any need for separate sheathing. Tests have shown that plywood as thin as $\frac{1}{4}$ inch nailed directly to studs provides twice the rigidity and nearly three times the strength of 1-inch lumber sheathing nailed to studs horizontally. Plywood is ideal for lightweight "stress skin construction." Gluing with mastic adhesive, in addition to nailing, provides even greater rigidity. Plywood siding comes in smooth or decorative surfaces, finished and unfinished.

Board siding is what most other siding materials imitate. Horizontal lap siding can be applied directly to studs, but diagonal bracing must first be let into the studs to prevent racking. Vertical board siding may also be applied directly to studs, with diagonal bracing, but filler blocks must be nailed between the studs so that siding can be nailed.

The siding is made in many forms and may be installed in a variety of ways to achieve desired appearances.

Hardboard siding is made in 48 × 96-inch panels, and 12-inch wide planks that are 8, 9 or 16 feet long depending on application. These can be installed the same way as board siding is. Panels may be applied directly to studs without sheathing or diagonal bracing.

SIDING INSTALLATION

Siding Material	Single Wall Construction		Over ⅜" Sheathing	
	Maximum stud spacing	Nail size and type	Maximum stud spacing	Nail size and type
Exterior Plywood				
⅜"	16"	6d noncorrosive	24"	6d noncorrosive
⅝" groove	16"	8d noncorrosive	24"	8d noncorrosive
⅝" flat	24"	8d noncorrosive	24"	8d noncorrosive
Lap siding	16"	8d noncorrosive	24"	8d noncorrosive
Board Siding				
Horizontal	16"	6d-8d noncorrosive	16"	6d-8d noncorrosive
Vertical	16"	6d-8d noncorrosive	16"	6d-8d noncorrosive
Hardboard Panel				
Shiplap edge	16"	6d-8d galvanized box	16"	6d-8d galvanized box
Butt edge	24"	6d-8d galvanized box	24'	6d-8d galvanized box
Hardboard Lap siding	16"	8d galvanized	16"	10d galvanized
Aluminum and Vinyl Lap Siding	No	No	24"	2" aluminum siding nails

STORAGE SHEDS FOR EVERY PURPOSE / 187

Hardboard lap siding requires either structural sheathing or diagonal bracing for racking strength.

Aluminum and vinyl horizontal lap and vertical board siding can also be used effectively on sheds and other yard structures. The siding should be hung over sheathing, but the framing/sheathing combination must provide all structural strength.

When estimating relative costs of the various sidings, be sure to figure in the cost of framing and bracing and/or sheathing that may be required. Plywood in general is the most economical siding.

Nail Spacing		
Edges	**Intermediate**	**Notes**
6″	12″	No building paper or corner bracing required.
6″	12″	
6″	12″	
16″ o.c.,	1″ from bottom	Use building paper. No corner bracing required.
Varies	Varies	
Varies	Varies	Block every 24″
4″	8″	Vapor barrier and caulk or battens required.
4″	8″	
Varies	Varies	Vapor barrier and metal outside corners required.
16″ spacing into studs if nonwood sheathing		Vapor barrier required.

PLYWOOD RUSTIC SURFACES

From top to bottom are board and batten, kerfed, texture 1-11, reverse board, batten-and-channel groove.

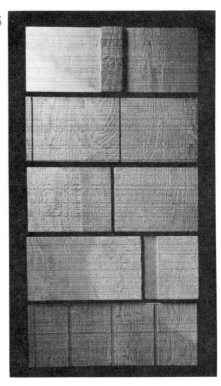

WOOD SIDING STYLES AND INSTALLATION

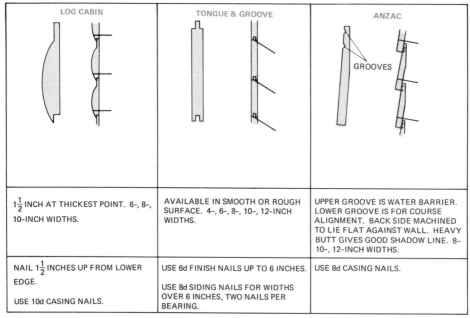

LOG CABIN	TONGUE & GROOVE	ANZAC
$1\frac{1}{2}$ INCH AT THICKEST POINT. 6-, 8-, 10-INCH WIDTHS.	AVAILABLE IN SMOOTH OR ROUGH SURFACE. 4-, 6-, 8-, 10-, 12-INCH WIDTHS.	UPPER GROOVE IS WATER BARRIER. LOWER GROOVE IS FOR COURSE ALIGNMENT. BACK SIDE MACHINED TO LIE FLAT AGAINST WALL. HEAVY BUTT GIVES GOOD SHADOW LINE. 8-10-, 12-INCH WIDTHS.
NAIL $1\frac{1}{2}$ INCHES UP FROM LOWER EDGE. USE 10d CASING NAILS.	USE 6d FINISH NAILS UP TO 6 INCHES. USE 8d SIDING NAILS FOR WIDTHS OVER 6 INCHES, TWO NAILS PER BEARING.	USE 8d CASING NAILS.

WOOD SIDING (Continued)

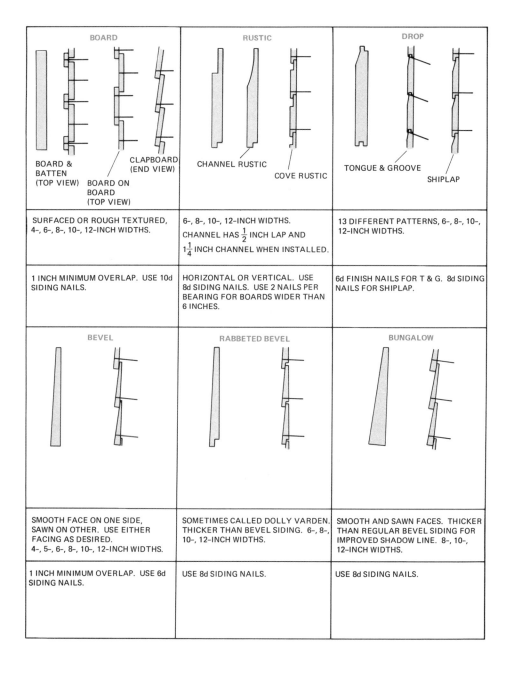

BOARD	RUSTIC	DROP
BOARD & BATTEN (TOP VIEW) BOARD ON BOARD (TOP VIEW) CLAPBOARD (END VIEW)	CHANNEL RUSTIC COVE RUSTIC	TONGUE & GROOVE SHIPLAP
SURFACED OR ROUGH TEXTURED, 4-, 6-, 8-, 10-, 12-INCH WIDTHS.	6-, 8-, 10-, 12-INCH WIDTHS. CHANNEL HAS $\frac{1}{2}$ INCH LAP AND $1\frac{1}{4}$ INCH CHANNEL WHEN INSTALLED.	13 DIFFERENT PATTERNS, 6-, 8-, 10-, 12-INCH WIDTHS.
1 INCH MINIMUM OVERLAP. USE 10d SIDING NAILS.	HORIZONTAL OR VERTICAL. USE 8d SIDING NAILS. USE 2 NAILS PER BEARING FOR BOARDS WIDER THAN 6 INCHES.	6d FINISH NAILS FOR T & G. 8d SIDING NAILS FOR SHIPLAP.
BEVEL	RABBETED BEVEL	BUNGALOW
SMOOTH FACE ON ONE SIDE, SAWN ON OTHER. USE EITHER FACING AS DESIRED. 4-, 5-, 6-, 8-, 10-, 12-INCH WIDTHS.	SOMETIMES CALLED DOLLY VARDEN. THICKER THAN BEVEL SIDING. 6-, 8-, 10-, 12-INCH WIDTHS.	SMOOTH AND SAWN FACES. THICKER THAN REGULAR BEVEL SIDING FOR IMPROVED SHADOW LINE. 8-, 10-, 12-INCH WIDTHS.
1 INCH MINIMUM OVERLAP. USE 6d SIDING NAILS.	USE 8d SIDING NAILS.	USE 8d SIDING NAILS.

LITTLE RED BARNS

Regardless of their color, miniature barns are a popular and practical storage shed. They are sturdy, and provide a lot of storage space for their cost and floor area. Some bigger models have a loft with a separate high door in addition to sliding and smaller side doors. Construction generally follows standard house building methods, but is of course scaled down.

A typical barn kit will include all framing members accurately cut to dimension and usually numbered so you know which is which. Also included will be plans and assembly instructions and a list of materials that you will have to buy locally. These materials would consist of

GARDEN-PLAY BARN

Here's a little barn with many possible uses, ranging from storage shed to children's playhouse. The front offers a Dutch door. The back features ramp-entry, double barn doors that open wide enough for a power tiller or garden tractor. Tools can be hung on Peg-Board hangers mounted on the sloping interior walls. There's a large storage loft with access under a hinged roof panel. Step-by-step instructions for construction are shown on the following pages. (*Plans courtesy of Masonite Corp.*)

sheathing and siding, shingles, roofing paper, wood and metal trim, and nails.

A prefabricated barn will arrive in large assemblies—front, back, sides, doors, and roof sections that bolt and nail together. Shingle trimming is sometimes necessary, and you have to provide the foundation.

Plans for barns are available from several sources. Using only plans, you will have to do some fairly accurate cutting on framing members with saber and portable circular saws. But this can save money over buying a kit. Plans also offer you the opportunity to modify a proven design to suit your needs, without the need to design the basic structure yourself. Then, using the plans as reference, you can make the barn longer or shorter, move or add doors and windows, or change the type of siding material or trim.

EXPLODED VIEW OF FRAMING

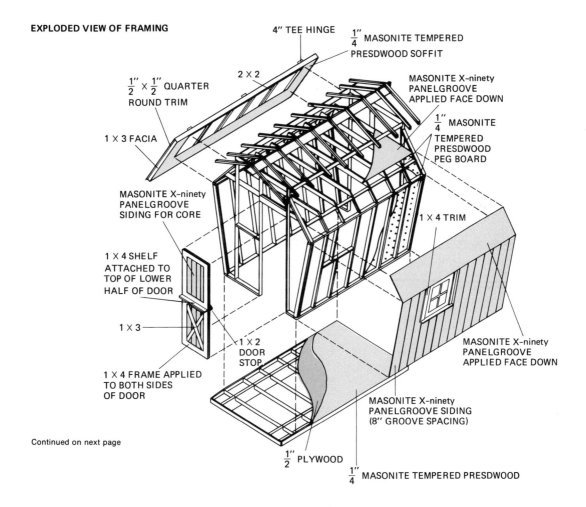

Continued on next page

EXPLODED VIEW OF FRAMING

GARDEN-PLAY BARN (*Continued*)

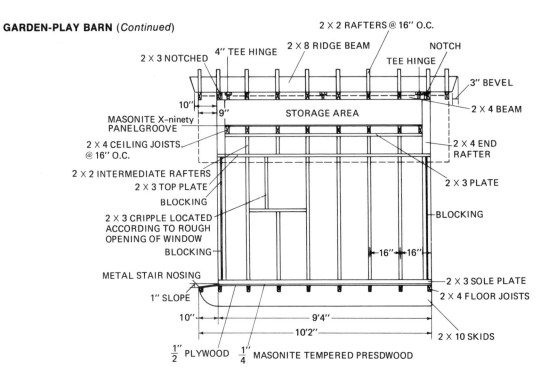

SIDE FRAMING

MATERIALS LIST AND BREAKDOWN

Materials List

	QUANTITY	SIZE
$\frac{1}{2}$" × $\frac{1}{2}$" quarter round	13	10'
1" × 3"	4	8'
	2	10'
	4	12'
1" × 4"	15	12'
2" × 2"	18	8'
	5	12'
2" × 3"	23	6'
	5	8'
	8	10'
2" × 4"	10	6'
	8	8'
	4	10'
	2	12'
2" × 8"	1	12'
2" × 10"	2	12'
Masonite Products:		
$\frac{1}{4}$" Tempered Presdwood	4	4' × 8'
X-ninety Panelgroove—	6	4' × 8'
8" Center	6	4' × 10'
$\frac{1}{4}$" Peg-Board	5	4' × 8'

	QUANTITY	SIZE
Plywood:		
$\frac{1}{2}$" plywood	2	4' × 8'
$\frac{5}{16}$" plywood	5	4' × 8'
Cedar shakes or		
asphalt shingles	2 bundles or 155 sq. ft.	
30 lb. building paper	155 sq. ft.	
Metal stair nosing	8'	
4" Tee hinges	14	
Spring lock	2	
Foot lock	3	
Square steel spring bolts	2	
4-lite barn-sash windows	3	2' × 2' 6"

MATERIALS BREAKDOWN

	QUANTITY	SIZE
FLOOR		
Joists	8	2" × 4" × 6'
	1	2" × 3" × 6'
Skids	2	2" × 10" × 12' (redwood, fir, or treated pine to weather)

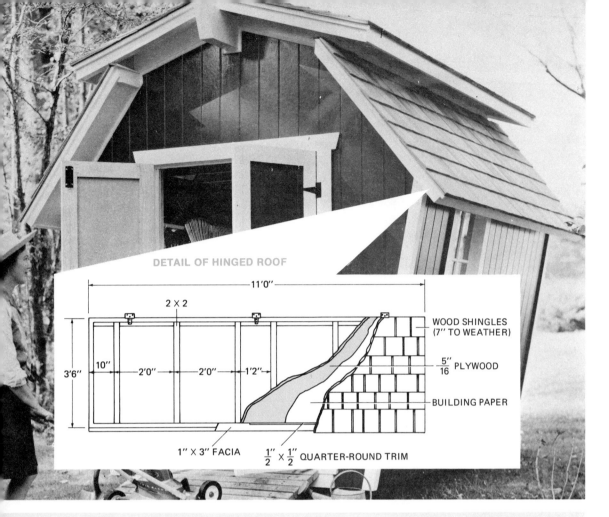

DETAIL OF HINGED ROOF

11'0''

2 X 2

3'6''

10'' 2'0'' 2'0'' 1'2''

WOOD SHINGLES
(7'' TO WEATHER)

$\frac{5''}{16}$ PLYWOOD

BUILDING PAPER

1'' X 3'' FACIA $\frac{1}{2}$'' X $\frac{1}{2}$'' QUARTER-ROUND TRIM

	QUANTITY	SIZE		QUANTITY	SIZE
Flooring	2	$\frac{1}{2}$'' × 4' × 8' plywood	Shingles		1 bundle cedar shakes, or 80 sq. ft. asphalt shingles
Finished Flooring	2	$\frac{1}{4}$'' × 4' × 8' Masonite Tempered Presdwood	Building Paper	80 sq. ft.	30 lb. felt
			Soffits	1	$\frac{1}{2}$'' × 4' × 8' Masonite Tempered Presdwood
Headers	2	2'' × 4'' × 12'			
Trim	8'	Metal stair nosing	Facia	2	1'' × 3'' × 12'
				4	1'' × 3'' × 4'
CEILING			Quarter round	4	$\frac{1}{2}$'' × $\frac{1}{2}$'' × 10'
Joists	8	2'' × 4'' × 8'	Hinges	6	4'' Steel Tee (extra heavy)
Ceiling panels	2	$\frac{7}{16}$'' × 4' × 10' Masonite X-ninety Panelgroove (8'' o.c. groove spacing)			
			STATIONARY PART OF ROOF		
			Ridge Beam	1	2'' × 8'' × 12'
			Rafters	20	2'' × 2'' × 4'
			Beams	2	2'' × 4'' × 10'
			Facia	2	1'' × 3'' × 12'
				4	1'' × 3'' × 4'
			Sheathing	2$\frac{1}{2}$	$\frac{7}{16}$'' × 4' × 8' Plywood
HINGED PART OF ROOF					
Framing	4	2'' × 2'' × 12'	Shingles		1 bundle cedar shakes, or 75 sq. ft. asphalt shingles
	16	2'' × 2'' × 4'			
Sheathing	2$\frac{1}{2}$	$\frac{5}{16}$'' × 4' × 8' plywood			

Continued on next page

GARDEN-PLAY BARN
(*Continued*)

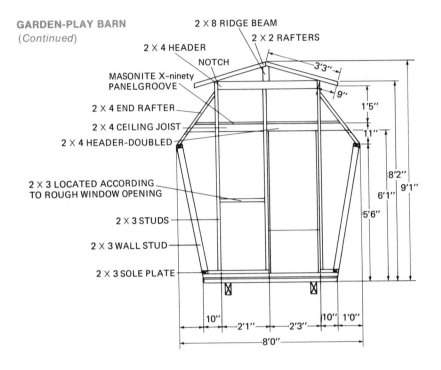

FRONT FRAMING

MATERIALS LIST (*Continued*)

	QUANTITY	SIZE	INTERIOR WALLS	QUANTITY	SIZE
Soffits	1	$\frac{1}{4}'' \times 4' \times 8'$ Masonite Tempered Presdwood		5	$\frac{1}{4}'' \times 4' \times 8'$ Masonite Tempered Presdwood Peg-board
Quarter Round Trim	4	$\frac{1}{2}'' \times \frac{1}{2}'' \times 10'$			
Building Paper	75 sq. ft.	30 lb. felt			
			EXTERIOR WALLS		
			Studs	5	$2'' \times 3'' \times 8'$
DOORS AND WINDOWS			Studs	22	$2'' \times 3'' \times 6'$
4-lite barn-sash windows	3	$2' \times 2' 6''$	Intermediate rafters	12	$2'' \times 2'' \times 1'$
Door core		$\frac{7}{16}''$ Masonite X-ninety Panelgroove (cut from door openings in end walls)	Sole and plates	6	$2'' \times 3'' \times 10'$
			End rafters	4	$2'' \times 4'' \times 3'$
			Miscellaneous (headers, stant, window blocking)	2	$2'' \times 3'' \times 10'$
				2	$2'' \times 4'' \times 10'$
			Siding	6	$\frac{7}{16}'' \times 4' \times 8'$ Masonite X-ninety Panelgroove
Trim	5	$1'' \times 4'' \times 12'$		4	$\frac{7}{16}'' \times 4' \times 10'$ Masonite X-ninety Panelgroove
	2	$1'' \times 3'' \times 10'$			
Hardware	8	4'' steel Tee Hinges (extra heavy)			
	2	Spring locks	Trim (doors, windows, corner boards)	10	$1'' \times 4'' \times 12'$
	3	Foot locks	Quarter round	5	$\frac{1}{2}'' \times \frac{1}{2}'' \times 10'$
	2	Square steel Spring bolts			

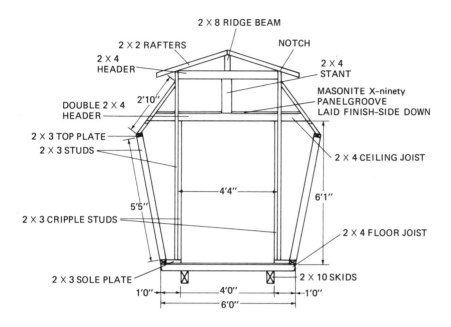

REAR FRAMING

GENERAL CONSTRUCTION HINTS

Application

Nails. When applying sidings directly to framework without sheathing or corner bracing, use only galvanized box nails.

Nail spacing. Space nails 6 inches apart at all vertical joints, 6 inches apart along horizontal edges, and 12 inches apart at intermediate framing. Nail $\frac{1}{2}$ inch from panel edges. Important! *Avoid nailing in the grooves.*

Joints. All joints and panel edges should fall opposite framing members. If it is necessary to make a joint with a panel that has been field cut and does not have a shiplap joint, use a butt joint. When making butt joints, butter edges with caulking and bring to light contact. Do not force or spring panels into place. Also apply caulking where panels are butted against window or door trim.

Painting

Primer. The factory applied primer is not a finish paint and must be painted within 60 days after installation. If siding is exposed for a longer period, reprime it with an exterior grade linseed oil base primer.

Finish Paint. For a gloss finish use a good quality conventional linseed oil base house paint. For a flat or low gloss finish use an alkyd type house paint formulated with linseed oil.

Water emulsion house paints (Latex types) generally require special primers or methods of application. Before using a water emulsion house paint read the manufacturer's instructions.

Number of coats. The number of coats required will depend on the method of application and paint used. Proper film thickness can best be obtained by applying two top coats over the primer. (*Continued on following pages.*)

GARDEN-PLAY BARN (*Continued*)

Step 1. Build the floor structure as shown in the photos here. First notch into the 2 × 10 skids for 2 × 4 floor joists. Make the notches at 16-inch spacings. Slope the ends of the skids for a sloped step that will run rain off. Bevel the bottom of the skid. (See Step 4 for cutting of pattern joists.)

Step 2. Space the skids 4 feet apart, measuring from the center of one skid to the center of the other. Level the skids and square them up before inserting floor joists. Leave joists uncut except for a pattern joist until all are in place; then use a chalk line to mark top edges of the remaining joists for cutting. Use a pattern or a set-guide for the bevel mark. Nail through the joists into skids with 8d coated box nails.

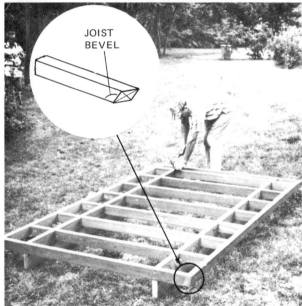

Step 3. Cover the floor framing with ½-inch plywood. Score the plywood on the backside as a guide for the bend you'll make for the sloped step. Cover the subflooring with ¼-inch Masonite Tempered Presdwood.

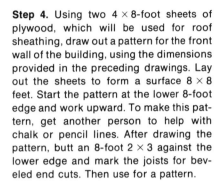

Step 4. Using two 4 × 8-foot sheets of plywood, which will be used for roof sheathing, draw out a pattern for the front wall of the building, using the dimensions provided in the preceding drawings. Lay out the sheets to form a surface 8 × 8 feet. Start the pattern at the lower 8-foot edge and work upward. To make this pattern, get another person to help with chalk or pencil lines. After drawing the pattern, butt an 8-foot 2 × 3 against the lower edge and mark the joists for beveled end cuts. Then use for a pattern.

STUDS BEVELED AT BOTTOM END

Step 5. Cut wall studs according to the pattern. Note that studs are beveled at the bottom but not at the top. Nail through top and bottom plates to frame an entire wall before tilting it into position. Make corner posts by separating two studs with blocking. Delay cutting the window opening until the wall frame is up and braced.

Step 6. After the sidewalls are up and braced, cut and assemble the 2 × 3 framing for end walls. Shown here is an assembly of a double 2 × 4 header for the front door. The header for rear double doors is made in the same fashion. After all pieces for the front wall are cut, use the same pattern, but making necessary changes to lay out the rear wall.

Continued on next page

DOUBLED HEADER

GARDEN-PLAY BARN (*Continued*)

Step 7. Raise end walls into position and join to sidewalls by attaching the diagonally placed 2 × 4 end rafters. Use 8d coated box nails or 10d for all rough framing. Use 16d common nails to nail through one member into another as in fastening plates to ends of studs.

Step 8. Install beams. Be certain end walls are completely plumb. Then measure between notches before cutting the two 2 × 4 side beams to length. Lay the ridge beam directly atop 2 × 4 headers, centered, and toe-nailed to the 2 × 4s.

Step 9. Next install the eight ceiling joists. First cut a pattern joist at the proper bevel. Then check it in position before cutting the remaining seven joists. Nail one joist in place on each end wall. Take measurements from those joists for the intermediate rafters that will support the remaining joists. Mark and rip (at a bevel) a 2 × 3 for joists to rest on. Cut 2 × 2 intermediate rafters to be placed between the longitudinal members and nail in place. Install the other ceiling joists.

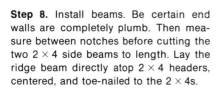

Step 10. Cut top roof rafters according to the large pattern (Step 4). If you've followed the pattern closely, these rafters will be 3 feet 4 inches long, allowing a 9-inch overhang. Make a pattern rafter with a notch at the point where it rests atop the 2 × 4 beam. Check the pattern rafter at various rafter positions before cutting the remaining rafters. Toenail rafters at the notch and at the ridge.

Step 11. Sheath and shingle the roof before starting to apply siding. Nail a 1 × 3 facia to the ends of the rafters so the facia extends a half-inch above and below the rafters. This will conceal the edges of the sheathing and the soffit. Continue the facia completely around the roof. Fasten the sheathing. Apply 30-pound building paper to the sheathing. When wood shingles are used on the roof, apply a double starter course that hangs ½-inch over the facia. Shingles should have a 5-inch exposure for each course. For the ridge, rip shingles to 3 inches wide and apply in overlapping pairs as shown. Nail ½ × ½-inch quarter-round trim to the facia directly under the bottom edge of the shingles.

Step 12. Apply the X-ninety Panelgroove Siding after framing rough window openings. To make the window openings, hold the window against the frame wall and mark for the opening. After you've framed the rough opening, hold the siding in position and mark the rough opening on the reverse side of the siding. Cut out the opening in the siding with a portable power saw, or start the opening with a keyhole saw and finish it with a regular crosscut saw.

Continued on next page

GARDEN-PLAY BARN (*Continued*)

Step 13. On sidewalls, there are few measuring problems. But on end walls, be sure to mark the outline for the siding according to the pattern. After cutting, put the panel in position and mark the outline for door cuts on the reverse side.

Step 14. Cover the ceiling with X-ninety Panelgroove laid grooved side down atop the ceiling joists. Then cover the intermediate rafters with the siding.

SIDING OVER INTERMEDIATE RAFTERS

Step 15. Build the lift-up roof sections according to the drawing showing the hinged roof. Use 2 × 2 framing spaced as shown. Frame and sheath the sections. Apply Tee-hinges over the sheathing. Then lift into place and attach the hinges to the 2 × 4 beams. Apply fascia, trim, and soffit as described earlier. Because of the steep slope, apply shingles with a 7-inch exposure.

RAKED TRIM

STAIR NOSING

Step 16. Trim the exterior. Note the raked top piece of trim over the door and window. A similar trim piece is atop the double doors on the rear. Use a metal stair nosing around the porch. Dress up the building by attaching quarter-round trim, wherever there are joints or where framing shows. Doors may be made of the pieces cut out from the siding for the rough openings. (See the exploded drawing at the beginning of these instructions).

Step 17. Attach 2 × 4 rafter over the corner post to form a nailing surface for the Peg-Board that will line the interior.

End of instructions for
Garden-play Barn

A "LIL" RED BARN
KIT ASSEMBLY

This is the assembled "Lil" Red Barn ready for paint. (*Photo sequence, courtesy of JER Manufacturing Co.*)

Step 1. After leveling the area where barn will stand, assemble the base on 4 × 4 skids. The barn can also use a concrete slab as base.

Step 2. Lay panels of textured siding on the flooring with tongue-and-grooved edges butted. Mark off the outline of the back of barn with templates provided and saw along the outline. Repeat for the front of the barn.

Step 3. Nail precut 1 × 6s to the butted front barn pieces; then saw out the doors to separate them from framed front. But leave part of the sawing till later so that the assembly holds together while you install the door hinges.

Step 4. Assemble the roof panels.

Step 5. Nail side walls to the base.

Step 6. After nailing the back to the base and side walls, nail the front to the base and side walls. Now finish sawing the doors apart.

Step 7. Erect the roof frame sections, attach roof boards, nail on gable and corner trim. And that's it.

9

Garden Work Centers, Large and Small

A garden work center can be simply a potting bench. At the other extreme, it can be an elaborate potting shed attached to a greenhouse and equipped for year-round gardening.

What do you need in a garden work center? To start, you will need a sturdy potting bench, high enough so that you can stand and work comfortably at it. Size is up to you. Whatever the surface, it should be able to stand up to rough usage and be easy to clean. The bench should be firmly anchored or else it might tip over when it is loaded with dirt-filled clay pots.

Peat moss, sand, loam, fertilizers, and other garden mixes require a dry area for storage. One way they can be stored is in metal or plastic garbage cans, and the cans can be stored tilted on racks to provide easier access.

Use the check list below as a basic guide for planning the storage area and layout of your own garden work center.

☐ Peat pots, flats
☐ Clay pots, sizes from $2\frac{1}{2}$ up to 10 inches
☐ Stakes, plant ties, and tags
☐ Notebook and pencils, waterproof marking pen
☐ Garden books
☐ Gloves
☐ Rakes, hoes, shovels, cultivators, etc.
☐ Small tools: trowel, cultivator, pruners, scissors, knife, measuring cups, and spoons
☐ Sieves

☐ Watering can
☐ Misting sprayer
☐ Containers to store soil, peat, sand, fertilizer, etc.
☐ Container for mixing
☐ Paper towels or newspapers
☐ Trash container
☐ Source of water
☐ Sink
☐ Shelves for potted plants, flats
☐ Insecticides, weed killers, etc. (These poisonous or harmful chemicals should be stored in a secure locked box.)

Built of rugged redwood, this gardening workshop includes special cupboards for long-handled tools, shelves for pots and small tools, and a work bench under which garbage cans fit. A sliding door locks in everything securely. Construction details are shown in the next chapter. (*Courtesy California Redwood Assn.*)

GARDEN KIOSK

This kiosk can serve simultaneously as a garden work center, a tool shed, and a greenhouse. The roof consists of two metal hoods, or awnings. All dimensions are controlled by hood dimensions. (*Photos courtesy California Redwood Assn. Plans courtesy Chevron Chemical Co., Ortho Div.*)

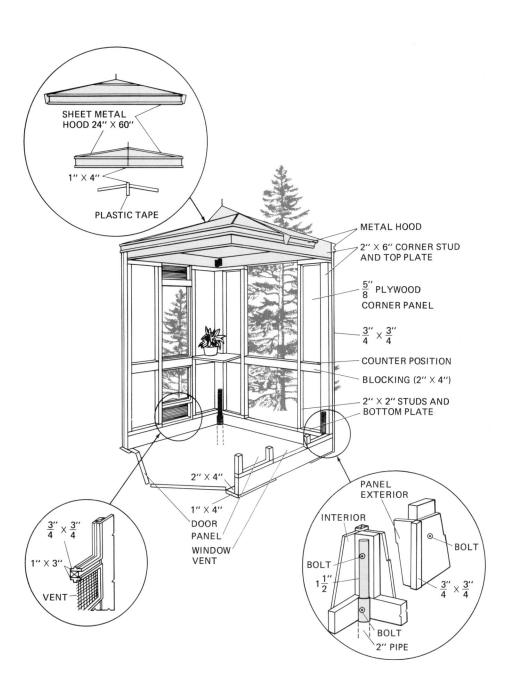

SHEET METAL
HOOD 24" X 60"

1" X 4"

PLASTIC TAPE

METAL HOOD

2" X 6" CORNER STUD
AND TOP PLATE

$\frac{5}{8}$" PLYWOOD
CORNER PANEL

$\frac{3}{4}$" X $\frac{3}{4}$"

COUNTER POSITION

BLOCKING (2" X 4")

2" X 2" STUDS AND
BOTTOM PLATE

2" X 4"

1" X 4"

DOOR
PANEL

WINDOW
VENT

$\frac{3}{4}$" X $\frac{3}{4}$"

1" X 3"

VENT

PANEL
EXTERIOR

INTERIOR

BOLT

BOLT

$1\frac{1}{2}$"

$\frac{3}{4}$" X $\frac{3}{4}$"

BOLT

2" PIPE

GARDEN LOCKER

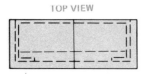

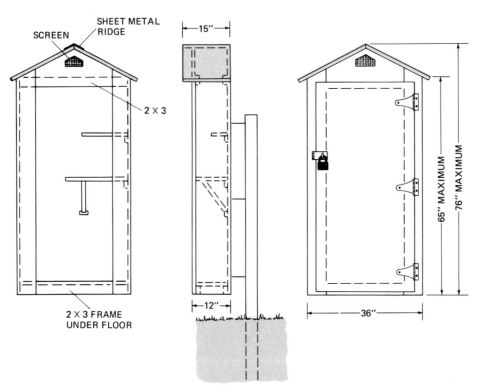

Mounted on a post, this design dates back to World War II victory gardens. For fun, neighbors might paint a crescent moon on the door. This structure can be made from two sheets of $\frac{5}{8}$-inch exterior textured plywood.

"NEWSSTAND" GARDEN CENTER

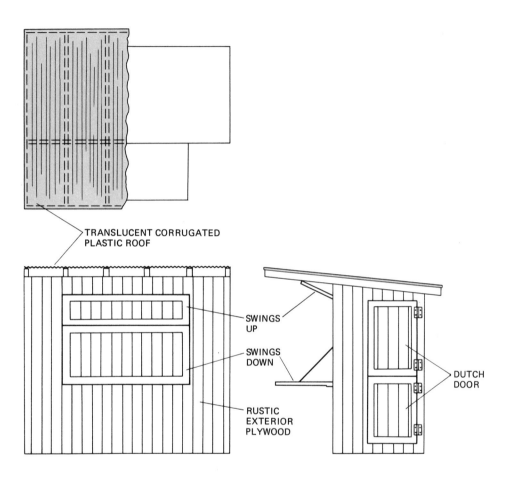

TRANSLUCENT CORRUGATED
PLASTIC ROOF

SWINGS
UP

SWINGS
DOWN

RUSTIC
EXTERIOR
PLYWOOD

DUTCH
DOOR

This garden work center resembles a big-city newsstand. It is conventionally framed with 2 × 3s and sheathed with rustic-textured exterior plywood. Its roof is made of translucent corrugated fiberglass reinforced plastic (FRP). There is a Dutch door—and a large swing-down work surface.

GARDEN CENTER

This garden work center has a tall, shallow tool storage section just inside the main doors, a large implement section reached through the side doors, a low greenhouse section in back, and a handy work counter on the porch. Detailed construction plans are shown on upcoming pages. (*Plan courtesy Masonite Corp.*)

APPROXIMATE BILL OF MATERIALS

CARPENTRY

No.	Size	Length	Description	No.	Size	Length	Description
3	2 × 4	15'	Base frame				
1	2 × 4	8'	Base frame	2	2 × 4	8'	Roof framing
21	2 × 4	8'	Deck	4	2 × 2	8'	Deck
3	2 × 4	6'	Vertical framing	8	1 × 1	4'	Window stops
5	2 × 4	5'	Vertical framing	2	1 × 1	8'	Door stops
3	2 × 4	3'	Vertical framing	21	2 × 2	6'	Door frames
2	2 × 4	8'6"	Roof brace	8	2 × 2	4'	Window frames
3	2 × 4	14'	Rafters	2	2 × 8	8'	Work shelf
2	2 × 4	10'	Rafters	1	1 × 4	8'	Trim
1	2 × 4	8'	Plate	1	1 × 6	8'	Trim

MASONITE PANEL PRODUCTS

No.				No.					
2	4' × 8' × ¼"	Masonite Tempered Presdwood	Floor	2	4' × 8' × ¼"		Masonite Weatherall	Door faces
3	4' × 8' × ¼"	Masonite Weatherall	Vertical walls	2	4' × 6' × ¼"		Masonite Tempered Peg-Board	Storage wall
				2	4' × 12' × 5/16"		Masonite Sunline Siding	Roof

MISCELLANEOUS

Door hardware, flashing strip, window glass, fasteners, paint.

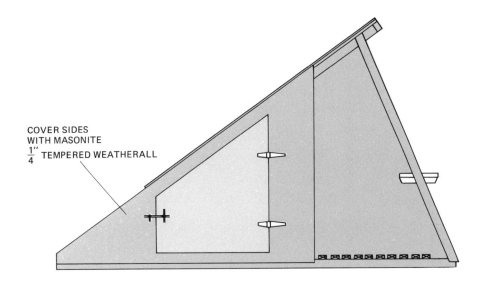

COVER SIDES
WITH MASONITE
$\frac{1''}{4}$ TEMPERED WEATHERALL

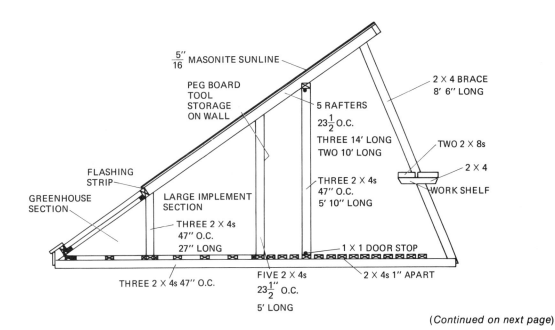

$\frac{5''}{16}$ MASONITE SUNLINE

PEG BOARD
TOOL
STORAGE
ON WALL

5 RAFTERS
$23\frac{1}{2}$ O.C.
THREE 14' LONG
TWO 10' LONG

THREE 2 X 4s
47'' O.C.
5' 10'' LONG

2 X 4 BRACE
8' 6'' LONG

TWO 2 X 8s

2 X 4
WORK SHELF

FLASHING
STRIP

GREENHOUSE
SECTION

LARGE IMPLEMENT
SECTION

THREE 2 X 4s
47'' O.C.
27'' LONG

1 X 1 DOOR STOP

THREE 2 X 4s 47'' O.C.

FIVE 2 X 4s
$23\frac{1}{2}$'' O.C.
5' LONG

2 X 4s 1'' APART

(Continued on next page)

GARDEN CENTER
(*Continued*)

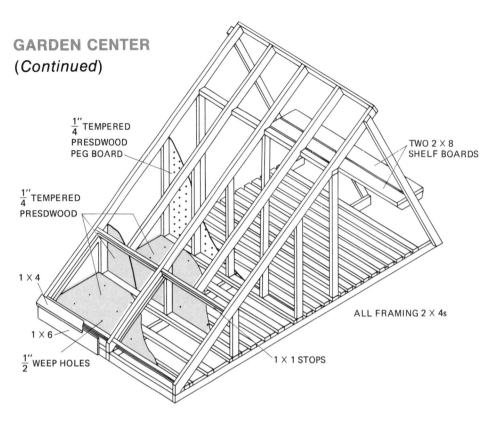

$\frac{1}{4}''$ TEMPERED PRESDWOOD PEG BOARD

$\frac{1}{4}''$ TEMPERED PRESDWOOD

1 X 4

1 X 6

$\frac{1}{2}''$ WEEP HOLES

TWO 2 X 8 SHELF BOARDS

ALL FRAMING 2 X 4s

1 X 1 STOPS

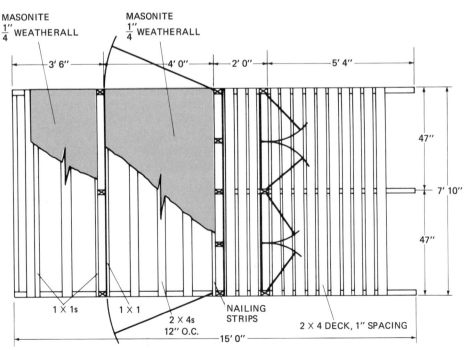

MASONITE $\frac{1}{4}''$ WEATHERALL

MASONITE $\frac{1}{4}''$ WEATHERALL

3' 6''

4' 0''

2' 0''

5' 4''

47''

7' 10''

47''

1 X 1s

1 X 1

2 X 4s 12'' O.C.

NAILING STRIPS

2 X 4 DECK, 1'' SPACING

15' 0''

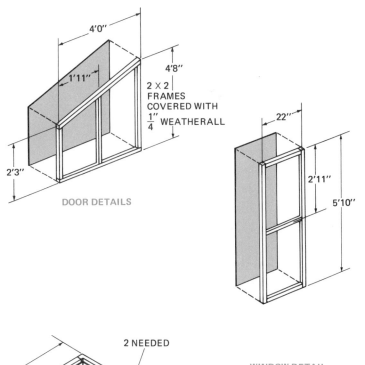

4'0''

1'11''

4'8''

2 X 2
FRAMES
COVERED WITH
$\frac{1}{4}$'' WEATHERALL

2'3''

DOOR DETAILS

22''

2'11''

5'10''

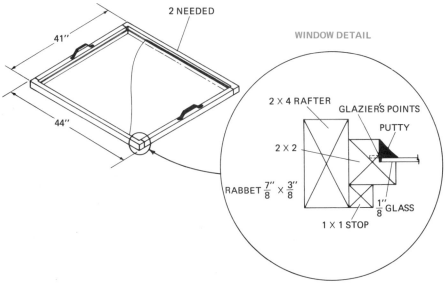

2 NEEDED

41''

44''

WINDOW DETAIL

2 X 4 RAFTER

GLAZIER'S POINTS

PUTTY

2 X 2

RABBET $\frac{7''}{8} \times \frac{3''}{8}$

$\frac{1}{8}''$ GLASS

1 X 1 STOP

10

Shelters with Unusual Effects

Gazebos are garden or yard shelters with a view. Storage sheds are structures primarily built for utility. In between the two is a wide array of mostly informal yard and garden structures whose main purpose is shelter. They can be used to provide shelter from the sun, from wind, from rain, from neighbors, from the street.

Shelters should be designed to be both practical and attractive. At one time, shelters were placed with geometric exactness in relation to the house. No more! It's an informal world today. The modern guideline is to put shelters where they will be useful. But when locating a shelter, you should consider both the view from the shelter, and the view of the shelter, particularly from the inside of your house.

Designs for shelters vary, mostly on the basis of what they are intended to shelter occupants from. For rain, a solid roof is required. For a prevailing seacoast wind, you'll need a solid wall. Shelter from the sun or from the second story windows of a neighbor's house can be achieved with a slatted, egg-crate bamboo, or a reed mat or other partially open roof. Privacy from the street is often best achieved by means of structure and foliage.

Shelters are often used with paved areas or terraces, and may not always cover the whole area. A shelter and a garden work center can often be combined to save space and cost.

Shelters usually contain seats and tables. The dimensions of chairs or benches for use at a table must show consideration for table

This redwood gazebo is designed for shade. But functionally it seems to be more a shelter than a gazebo. The high-pitched slatted roof is set square with the base. The side wall frame is set diagonally on the base. Louvered sides can be positioned for maximum shade. (*California Redwood Assn*.)

This poolside shelter of redwood could be modified to provide dressing rooms that might occupy only part of the floor area. (*Courtesy California Redwood Assn*.)

postures. Benches for relaxation can be more varied in dimension. The size and height of table tops depends on the uses the table will serve.

Unless a shelter is to be strictly a temporary structure, such as to shade some young trees, it should be constructed with care, using the best materials.

You'll find that much of the information on construction in earlier chapters applies for the construction of shelters. So if this chapter leaves some questions unanswered, you'll probably find what you need in the construction details of the earlier chapters.

This A-frame shelter has redwood 2 × 6 rafters, a 4 × 6 ridge, and 2 × 2 slats. Bottom ends of rafters must be firmly anchored to prevent spreading. (*Courtesy California Redwood Assn.*)

A canvas canopy makes a lightweight and inexpensive shade cover for this hillside deck. Perforated metal cylinders at the left and right sides conceal lighting. (*Courtesy California Redwood Assn.*)

Here a redwood pergola provides decorative shade for a gravel walk. (*Courtesy California Redwood Assn.*)

A shade shelter for a greenhouse provides additional growing space, and helps control the temperature inside the greenhouse. (*Courtesy California Redwood Assn.*)

REDWOOD GARDEN SHELTER

This complex garden shelter functions as a gazebo, a garden work center and a storage shed, all in one. The design contains many features that can be used in smaller shelters. (*Courtesy California Redwood Assn.*)

FRONT VIEW

BACK VIEW

TOP VIEW

Here are the major lumber components of the garden shelter.

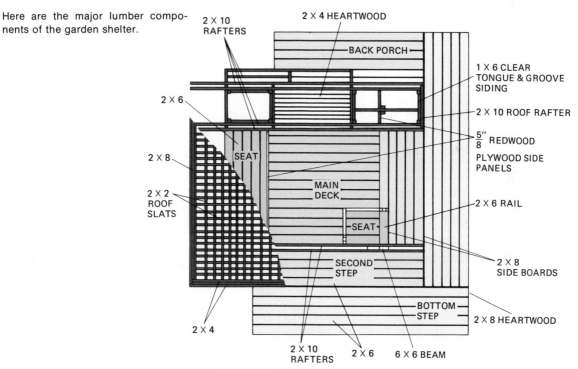

2 × 10 RAFTERS

2 × 4 HEARTWOOD

BACK PORCH

1 × 6 CLEAR TONGUE & GROOVE SIDING

2 × 6

2 × 10 ROOF RAFTER

$\frac{5}{8}''$ REDWOOD

PLYWOOD SIDE PANELS

SEAT

2 × 8

MAIN DECK

2 × 2 ROOF SLATS

2 × 6 RAIL

SEAT

2 × 8 SIDE BOARDS

SECOND STEP

BOTTOM STEP

2 × 8 HEARTWOOD

2 × 4

2 × 10 RAFTERS

2 × 6

6 × 6 BEAM

JOISTS AND SUPPORTING DECK 2 × 8 HEARTWOOD

A shelter with a solid, redwood board roof provides a shady entertaining area. The raised floor sets the shelter apart from the garden area. (*Courtesy California Redwood Assn.*)

A platformed shelter can provide late-afternoon shading for entertaining. Here 6 × 6 redwood posts support open roof construction. (*Courtesy California Redwood Assn.*)

This entrance pergola provides shelter and an air of informality. The floating appearance of the steps gives a lightness to the massive redwood timbers overhead. (*Courtesy California Redwood Assn.*)

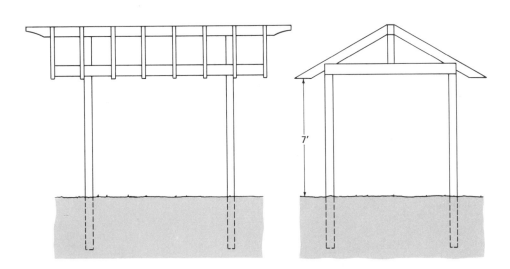

A simple framed shelter like this can be covered in many ways.

SHELTER WITH FOLDED ROOF

A folded-plate roof can be slatted, as shown, or covered with exterior plywood. This shelter roof is simple and inexpensive to build. (*Courtesy Louisiana-Pacific*)

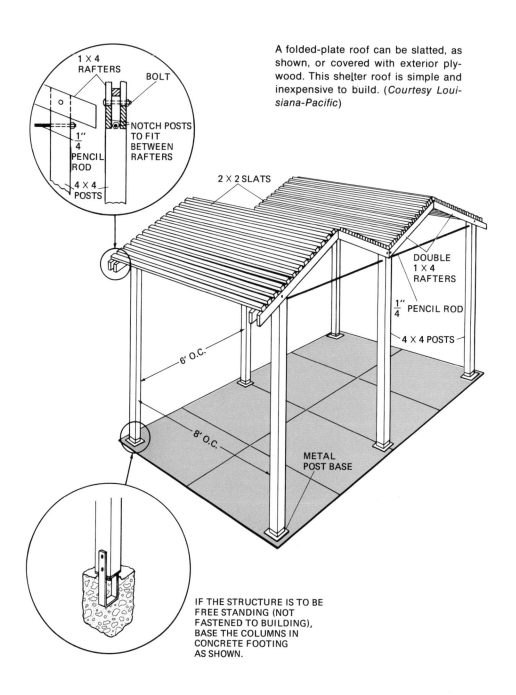

1 X 4 RAFTERS

BOLT

NOTCH POSTS TO FIT BETWEEN RAFTERS

$\frac{1}{4}$" PENCIL ROD

4 X 4 POSTS

2 X 2 SLATS

DOUBLE 1 X 4 RAFTERS

$\frac{1}{4}$" PENCIL ROD

4 X 4 POSTS

6' O.C.

8' O.C.

METAL POST BASE

IF THE STRUCTURE IS TO BE FREE STANDING (NOT FASTENED TO BUILDING), BASE THE COLUMNS IN CONCRETE FOOTING AS SHOWN.

STOW-IT-ALL FENCE or SHED

This Stow-It-All Fence or Shed is made of five 3-foot modules set side by side on 4 × 6 wood rails. The modules can be attached to vertical supports so that the units are really a shed. All doors have storage shelves built inside the frames. Siding may be butted or overlapped. *Modules 1 and 2* open from the left end and have a drop-down ramp for a lawnmower or other large item. The lower half of Module 1 opens into 2. *Module 3* contains shelving. *Module 4* is designed for storing long items, with appropriate hooks and hangers. *Module 5* has space below for large light items and lockable cabinets above for insecticides. (*Plans adapted from originals by Western Wood Products Assn.*)

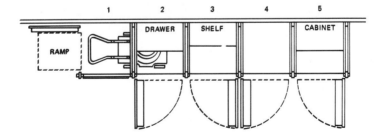

(*Continued on upcoming pages*)

STOW-IT-ALL FENCE (*Continued*)

FRONT VIEW

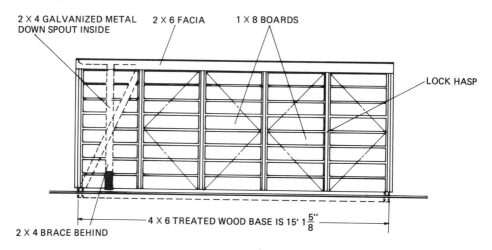

2 X 4 GALVANIZED METAL DOWN SPOUT INSIDE

2 X 6 FACIA

1 X 8 BOARDS

LOCK HASP

4 X 6 TREATED WOOD BASE IS 15' 1$\frac{5}{8}$"

2 X 4 BRACE BEHIND

TOP VIEW

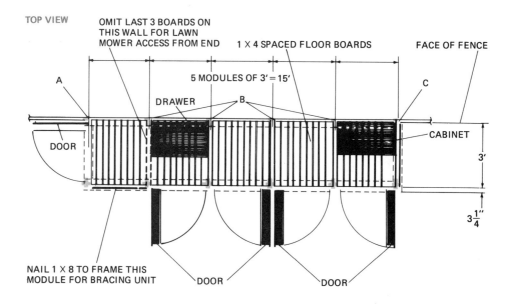

OMIT LAST 3 BOARDS ON THIS WALL FOR LAWN MOWER ACCESS FROM END

1 X 4 SPACED FLOOR BOARDS

FACE OF FENCE

A

5 MODULES OF 3' = 15'

C

DRAWER

B

DOOR

CABINET

3'

3$\frac{1}{4}$"

NAIL 1 X 8 TO FRAME THIS MODULE FOR BRACING UNIT

DOOR

DOOR

DETAILS OF TOP VIEW

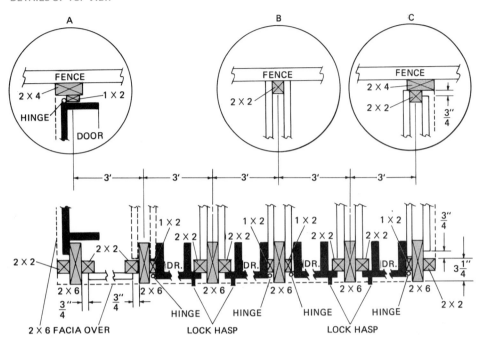

A

FENCE

2 X 4

1 X 2

HINGE

DOOR

B

FENCE

2 X 2

C

FENCE

2 X 4

2 X 2

3"/4

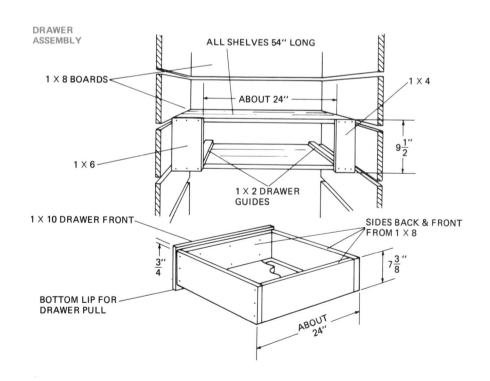

DRAWER ASSEMBLY

ALL SHELVES 54" LONG

1 X 8 BOARDS

ABOUT 24"

1 X 4

1 X 6

9 1/2"

1 X 2 DRAWER GUIDES

1 X 10 DRAWER FRONT

3"/4

SIDES BACK & FRONT FROM 1 X 8

7 3/8"

BOTTOM LIP FOR DRAWER PULL

ABOUT 24"

STOW-IT-ALL FENCE (*Continued*)

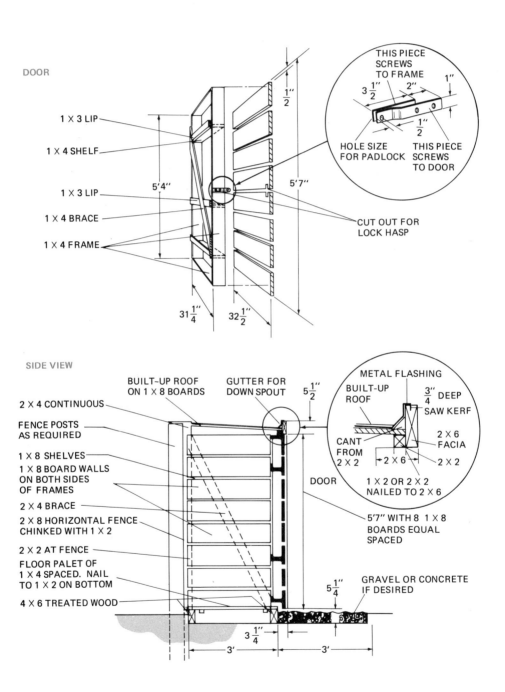

DOOR

THIS PIECE
SCREWS
TO FRAME

$3\frac{1}{2}$ 2″   1″

$\frac{1}{2}$″

HOLE SIZE
FOR PADLOCK

THIS PIECE
SCREWS
TO DOOR

1 X 3 LIP

1 X 4 SHELF

1 X 3 LIP

1 X 4 BRACE

1 X 4 FRAME

$\frac{1}{2}$″

5′4″

5′7″

CUT OUT FOR
LOCK HASP

$31\frac{1}{4}$″

$32\frac{1}{2}$″

SIDE VIEW

BUILT-UP ROOF
ON 1 X 8 BOARDS

GUTTER FOR
DOWN SPOUT

$5\frac{1}{2}$″

METAL FLASHING

BUILT-UP
ROOF

$\frac{3}{4}$″ DEEP
SAW KERF

2 X 4 CONTINUOUS

FENCE POSTS
AS REQUIRED

1 X 8 SHELVES

1 X 8 BOARD WALLS
ON BOTH SIDES
OF FRAMES

2 X 4 BRACE

2 X 8 HORIZONTAL FENCE
CHINKED WITH 1 X 2

2 X 2 AT FENCE

FLOOR PALET OF
1 X 4 SPACED. NAIL
TO 1 X 2 ON BOTTOM

4 X 6 TREATED WOOD

CANT
FROM
2 X 2

2 X 6

2 X 6
FACIA

2 X 2

1 X 2 OR 2 X 2
NAILED TO 2 X 6

DOOR

5′7″ WITH 8 1 X 8
BOARDS EQUAL
SPACED

GRAVEL OR CONCRETE
IF DESIRED

$5\frac{1}{4}$″

$3\frac{1}{4}$″

3′

3′

CABINET

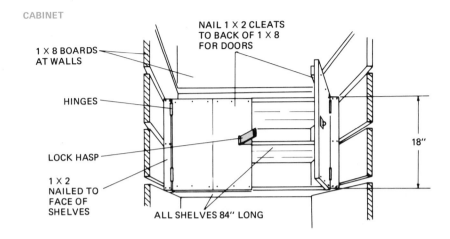

NAIL 1 × 2 CLEATS
TO BACK OF 1 × 8
FOR DOORS

1 × 8 BOARDS
AT WALLS

HINGES

LOCK HASP

1 × 2
NAILED TO
FACE OF
SHELVES

18"

ALL SHELVES 84" LONG

MATERIALS LIST FOR SINGLE MODULE			LINEAL FEET
Door			
Lip	1 × 3		8
Shelf (sides			
& brace)	1 × 4		30
Front	1 × 8	(32½" LONG × 8 = 21' 8")	22
Walls	1 × 8		144
Mullions	2 × 6		24
Mullions	2 × 2		42
Mullions	1 × 2		6
Braces	2 × 4		18
Flooring	1 × 4	9 pieces 36" long	27
Nailer	2 × 2		3
Base	4 × 6	Pressure treated	12
OR TWO 2 × 6 Nail-laminated			

Roof			LINEAL FEET
Sheath	1 × 8		15
Nailer	2 × 2		3
Nailer	2 × 4		3
Facia	2 × 6		14

For additional material for interior shelves and drawers, see materials list for entire unit.
- Pair of 3½" butt hinges
- 1 latch
- 28 gauge sheet metal flash as required

MATERIALS LIST OF 5 UNIT			LINEAL FEET
Doors			
Shelf lip	1 × 3	8 ft. × 5	40
Shelf & brace	1 × 4	30 ft. × 5	150
Front	1 × 8	22 ft. × 5	110
Walls	1 × 8		360 (*120)
	2 × 6		72 (*36)
Mullion	2 × 2		114 (*36)
	1 × 2		30
Braces	2 × 4		60 (*30)
Partition at			
Back	2 × 4	+ 12 ft. at wall or fence	+ 12
Roof			
Sheath	1 × 8		72
Nailer	2 × 2		16
Nailer	2 × 4		16
Facia	2 × 6		39 (*16)
Flooring	1 × 4	27 ft. × 5 = 135 + 2 extra 3' pieces	141
Nailer	2 × 2		30
Base	4 × 6	Pressure Treated	38

Interior			LINEAL FEET
Storage Units			
Shelves	1 × 8	*4 Boards in each module 36" long	60
Drawer	1 × 8	Sides 8 ft. bottom 6 ft.	14
(each)	1 × 10	Drawer front	2
	1 × 2	Side glides	4
Cabinet	1 × 8	Front	6
Front (each)	1 × 2	Styles	9

*Indicates material not required when unit backs up to fence or wall
Miscellaneous Hardware
- 45" × 1" × ⅞" galvanized steel cut as shown for the lock clasp
- 5 pair 3½" butt hinges for main doors
- 2 pair 2" butt hinges for locked cabinet
- 5" × 36' 28 gauge galvanized flashing for roof
- 2" × 4" × 6' galvanized metal down spout and leader

This list does not include fence materials. Instead it allows for braces and vertical supports at the back wall that could replace a fence.

Appendix :
Sources for Information and Materials

GREENHOUSE KIT MANUFACTURERS

Aluminum Greenhouses, Inc., 14615 Lorain Ave., Cleveland, OH 44111

Baco Leisure Products, Inc., 19 East 47th St., New York, NY 10017

W. Atlee Burpee Co., Warminster, PA 18974

Casaplanta, 16129 Cohasset St., Van Nuys, CA 91406

Clover Garden Products, 2596 Bransford Ave., Nashville, TN 37204

Dome East Corp., 325 Duffy Ave., Hicksville, NY 11801

Grow House Corp., 2335 Burbank St., Dallas, TX 75235

J. A. Nearing Co., Inc. (Janco), 10788 Tucker St., Beltsville, MD 20705

Lord & Burnham, Irvington-on-Hudson, NY 10533

McGregor Greenhouses, 1195 Thompson Ave., Santa Cruz, CA 95063

National Greenhouse Co., Pana, IL 62557

Redwood Domes, PO Box 666, Aptos, CA 95003

Peter Reimuller, 980—17th Ave., PO Box 2666, Santa Cruz, CA 95063

Rough Brothers, PO Box 16010, Cincinnati, OH 45216

Solar Technology Corp., 2160 Clay St., Denver, CO 80211

Sturdi-built Manufacturing Co., 11304 SW Boones Ferry Rd., Portland, OR 97219

Sun America Corp., PO Box 125, Houston, TX 77001

Texas Greenhouse Co., Inc., 2717 St. Louis, Fort Worth, TX 76110

Turner Greenhouses, PO Box 1260, Highway 117 South, Goldsboro, NC 27530

Vegetable Factory, Inc., 100 Court St., Copiague, Long Island, NY 11726

SOURCES OF KITS, PREFABS, AND PLANS

Wood Kits and Prefabs

Jim Dalton Garden House Co., 906 Cottman Ave., Philadelphia, PA 19111 (Prefab sheds, barns, small houses)

Jer Manufacturing Co., 280 River, Coopersville, MI 49404 (Precut kits for barns, several styles, sizes)

Stoltzfus Structures, R. D. 1, Kinzers, PA 17535 (Prefab barns, several styles)

Walpole Woodworkers, Inc., 767 East St., Walpole, MA 02081 (Prefab sheds in many styles)

Metal Sheds

Arrow Group Industries, 100 Alexander Ave., Pompton Plains, NJ 07444 (Gable and gambrel roofs)

Eastern Products Corp., 9325 Snowden River Parkway, Columbia, MD 21406 (Gable roofs)

Robco, 950 Summit St., Niles, OH 44446 (Gable, mansard and gambrel roofs)

Wheeling Corrugating Co., Wheeling, WV 26003 (Gable and gambrel roofs)

Plans

American Plywood Association, 1119 A St., Tacoma, WA 98401

Arcadia Sheds, 82 Arcadia Rd., Westwood, MA 02090

Better Homes and Gardens Project Plans, Des Moines, IA 50336

California Redwood Association, 1050 Battery St., San Francisco, CA 94111

Louisiana-Pacific, 1300 SW Fifth Ave., Portland, OR 97201

Mechanix Illustrated, Fawcett Bldg., Greenwich, CT 06830

Popular Mechanics, 224 West 57th St., New York, NY 10019

Popular Science, Plans Dept., 380 Madison Ave., New York, NY 10017

Woman's Day, Fawcett Bldg., Greenwich, CT 06830

Index